How to Understand the Reading of Blue Print Drawings

The Fundamental Principles Simply Explained

Edward R. Vigneau

SOTHIS PRESS

ISBN 978-1-0368-0563-0

This edition is published by Sothis Press © 2023. The company is registered in England and Wales. The Company Number is 14900389.

HOW TO UNDERSTAND *the Reading of* BLUE PRINT DRAWINGS

THE FUNDAMENTAL PRINCIPLES SIMPLY EXPLAINED

BY

EDWARD R. VIGNEAU, I. E.

Author of Theory of Operation, Vaporization, Ignition Care and Management of Two-Cycle Gasoline Engines.

Member of Detroit Engineering Society.

---

228 ILLUSTRATIONS

All illustrations made by the author.

PRICE $2.00

PUBLISHED BY

EDUCATIONAL INSTITUTE

100 ROWLAND BLDG., DETROIT, MICHIGAN.

CONTENTS

PART I

THE ELEMENTS OF BLUE PRINT DRAWINGS

CONTENTS

PART VII

ASSEMBLIES

PART VIII

MISCELLANEOUS IN BLUE PRINT DRAWINGS

PART IX

CONVENTIONS

INTRODUCTION

Realizing the need of a book of this nature, the author has endeavored to place before mechanics and others who are in need of instruction on the reading of blue print drawings, a book that contains a clear and comprehensive discussion in most simple terms.

Throughout this book, I have used simple objects as illustrations upon which to base the instructions.

Knowing that the weakness of many a writer of educational texts, lies in his failure to make known the things that are known to him as very simple, though in reality very hard to the reader, I have tried to mention such simple facts in my explanation as would prove to be a stumbling block if not understood by the student. I have endeavored to make the subjects interesting and appealing, so as to leave no doubt in the reader's mind of every step that would lead to a complete grasp of reading a blue print drawing.

I do not wish to convey the idea that by studying this book, that the student can read a blue print drawing with the ease of a man who has given earnest study with long practical experience. I have not tried to teach the most complicated of blue print drawings, but to give the student fundamental principles, which with earnest efforts he may apply in practice.

I am gratefully indebted to Mr. G. J. Lux of the Studebaker Corporation of this city for valuable suggestions and criticisms extended me for guidance while working on this manuscript.

EDWARD R. VIGNEAU.

Detroit Michigan, January 25, 1919.

KEY TO READING.

In order that the reader may follow the subject more readily, the following type styles have been adopted throughout the book:

Bold face type is used for emphasis, and in refering to the chapter subject.

Name of the object referred to, begins with Capital letter.

CHAPTER TITLES, AND WORDS TAKEN FROM THE BLUE PRINT DRAWINGS, ARE SET ALL IN CAPITAL LETTERS.

PART I

PERSPECTIVE DRAWINGS

To begin with, look at Fig. 1, which is a **perspective view** drawing of a box.

A **perspective view** of an article is a representation of the article as it appears actually to our eyes. Many parts of the article thus appear in a single representation. The **perspective drawing** in this respect differs from the blue print drawing or the working drawing, which shows each part in separate detail.

Figure 1, then, is not a working drawing, but a **perspective drawing,** as the foregoing explanation makes clear.

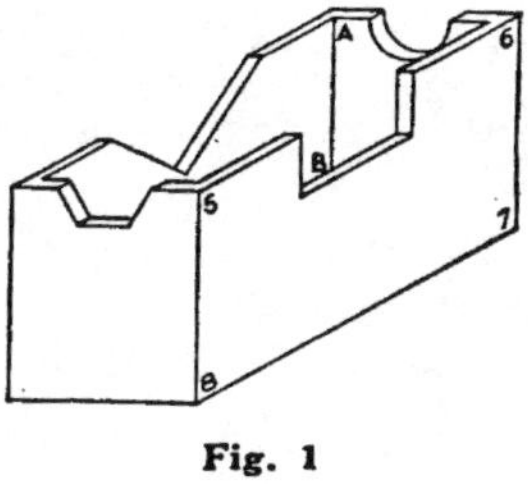

Fig. 1

In Fig. 1 the **perspective drawing** shows the ends and the sides of the Box is cut out in various shapes. The **perspective drawing** pictures the Box as it might perhaps appear in reality.

As you look at the Box as represented in Fig. 1, you see three full views that form the right side, the front end, and top views, but you cannot see any of the bottom, or all of the left side and back end. The right side and the front end views are partly in front of them.

In Fig. 2, which is the same Box as Fig. 1, three dotted lines have been placed.

DOTTED LINES

Dotted lines, it may be said just here, are used in a blue print drawing to represent **hidden lines.** The **dotted lines** in Fig. 2, in part, outlines the bottom of the Box, as it would look in Fig. 1 if you could see through the right side and front end.

In Fig. 2 the line from "A" to "C" can be only partly seen, because it is partly drawn full and partly dotted, which shows that the line from "A" to "B" can be seen above the side portion of the Box, but that the rest of the line which is from "B" to "C" cannot be seen, although it is understood as seen. Therefore, **hidden lines** to be understood as seen **must be dotted lines.** The last sentence is one of the general principles in blue print drawings.

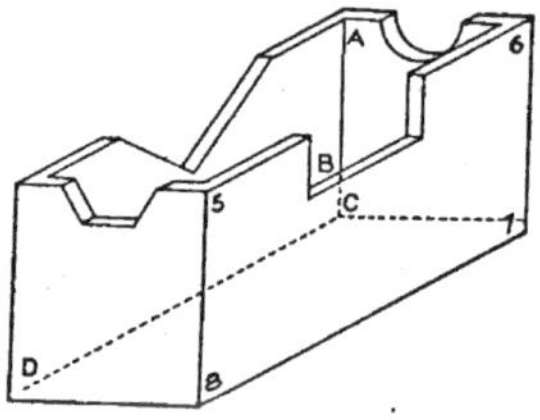

Fig. 2

All **dotted lines** on a blue print drawing have something in front of them. When reading a blue print drawing, see what these **dotted lines** form. These **dotted lines** represent what lies back of the outside of the object drawn.

To read understandingly the chapters that follow in this book, it is necessary for you to know exactly what **dotted lines** mean in a blue print drawing, for the meaning of **dotted lines** is one of the most important things to know for an understanding of blue print drawings.

FULL LINES

Full lines of any view drawing on a blue print drawing will always show you the outside shape and form for that view.

Full lines seen in any blue print drawing represent parts openly seen without any surfaces standing in front of them.

Figures 1 and 2 show the use of full and **dotted** (hidden) lines.

PLANE

The term "**plane**" that is to be understood in the reading of blue print drawings is no less important than that of dotted lines.

Let imagine a perfectly flat surface of glass, in thickness even thinner than a sheet of paper, is for all practical purposes to be considered a **plane.** In Fig. 3, let us think of the sheet, 1-2-3-4 as such a surface.

VIEWS

When a blue print drawing is made of any article, each side of the article as top, bottom, side or end is usually made into a separate **view** drawing, and the several **view** drawings are grouped around each other on a plane, so as to make a complete blue print drawing of the whole article.

Each separate **view** drawing shown on a plane is commonly called a "**view.**" By looking at each **view,** and studying each **view** at one time, a complete understanding may be had of each and every part that can be seen in certain sides of the article that the blue print drawing may show.

It is necessary that not only each side of the article represented be drawn on one plane as **views,** but that each **view** be grouped around one common **view** in such a manner that the relative positions of each side of the article drawn may be known as its **view** from the other **views** shown in a blue print drawing.

PART II

STRAIGHT VIEW PROJECTION

The word "**projection**" as applied to views, means that several sides or surfaces of an object are represented point for point on a straight surface known as a plane.

There are **straight** and **angle view projections,** both of which must be understood in the reading of blue print drawings. The **straight view projection** is that which is **projected** upon a plane in a straight to the edge of the drawing position. The **angle view projection** will not be discussed at this time, because it will be much more readily understood after the **straight view projection** has been studied.

In Fig. 3, you have represented the same Box as shown in Fig. 2 with the side 5-6-7-8 of the Box to lie in the plane 1-2-3-4. This plane is to be thought of as a transparent sheet of paper through which may be seen each move that the sides of the Box make as they swing into view upon the plane 1-2-3-4. This plane as has been stated, is in an upright slanted position. See Fig. 3.

The side 5-6-7-8 of the Box of Fig. 1 is placed against the plane 1-2-3-4 of Fig. 3 with the Box itself projecting to the rear of the plane, leaving only the side 5-6-7-8 of the Box to be actually seen on the plane as illustrated in Fig. 3.

Figure 3 shows only one view, for the other sides of the Box are not to be thought of as seen in the blue print drawing plane, unless these sides are swung around just enough to lie in the plane. It is only when the sides have thus been swung around, giving them position in the plane, that they become known as views. For convenience of illustration, hinges are to be imagined fastened to the Box as shown in Fig. 3.

More than one view of the Box may be obtained as illustrated in Fig. 4.

All sides of the Box of Fig. 4 may be imagined as swung on the hinges until the five sides lie against the plane 1-2-3-4 and are seen as five views. Particular stress must be given to your study at this point.

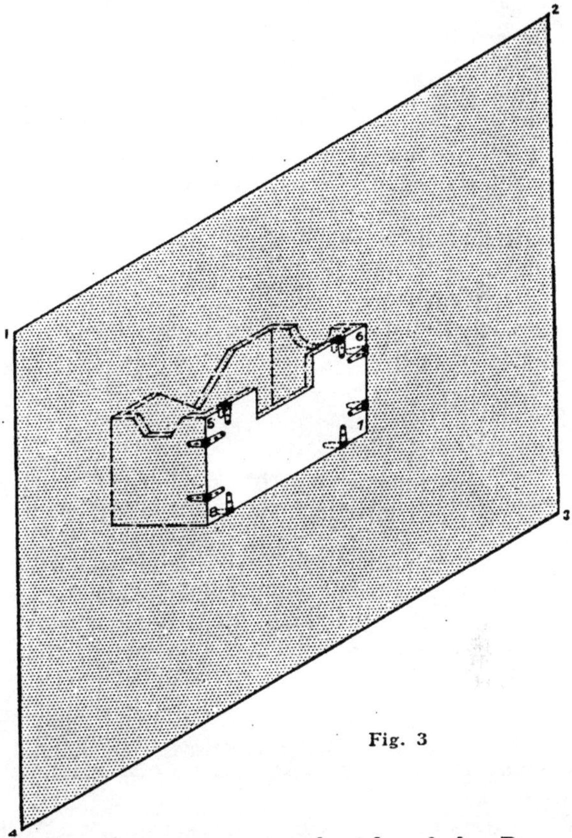

Fig. 3

Imagine that you see each side of the Box of Fig. 4, swing into the plane from the original position as shown in Fig. 3.

A very important point to be borne always in mind when reading a blue print drawing in **straight**

view projection is that the views are kept in the same relative position as the part of the object that they represent. Note in Fig. 4, that the sides of the Box that are hinged to the side 5-6-7-8 are swung into the plane from the position occupied in Fig. 3, are the same sides of the Box as hinged to the side 5-6-7-8 that is in the plane 1-2-3-4 of Fig. 4.

The sides of the Box which swing into the plane and become views should always be thought of when reading a blue print drawing as swinging into the plane in the direction of the opening hinges. Each **view projection** that swings into position as another view to a plane, is that which existed between the several sides in their original position in Fig. 3.

With the sides of the Box all hinged to the plane 1-2-3-4, as shown in Fig. 4, so that all of the five sides lie in the plane, each view drawing is then called a **straight view projection** from the view that it is next to. When this is done, all sides of the Box are shown as views on a plane.

The complete layout of views as seen in Fig. 4 is made to show the relation of each side of the Box to the other sides. Each side is a separate view only when the sides of the object drawn are shown on the plane.

The above explanation may seem so simple to you that you may question the wisdom of giving so much attention to the way in which the sides of the Box are swung to the plane 1-2-3-4 as so many views to be studied in the drawing. It may be said that the Box was selected as a highly simplified example to which to apply these principles that will aid you later in understanding the **projection** of objects much harder to understand.

Consider, for example, the Ratchet blue print drawing of Figs. 48 and 50, and you will realize the importance of the **"hinge method"** as an aid to understanding the way in which the end view is **projected** from the side view.

In order that you may become skillful in reading with clear understanding the relation between views in a blue print drawing, you must carefully master the explanation given with regard to Fig. 4 so as to imagine the hinges as strapped to the sides of the Box

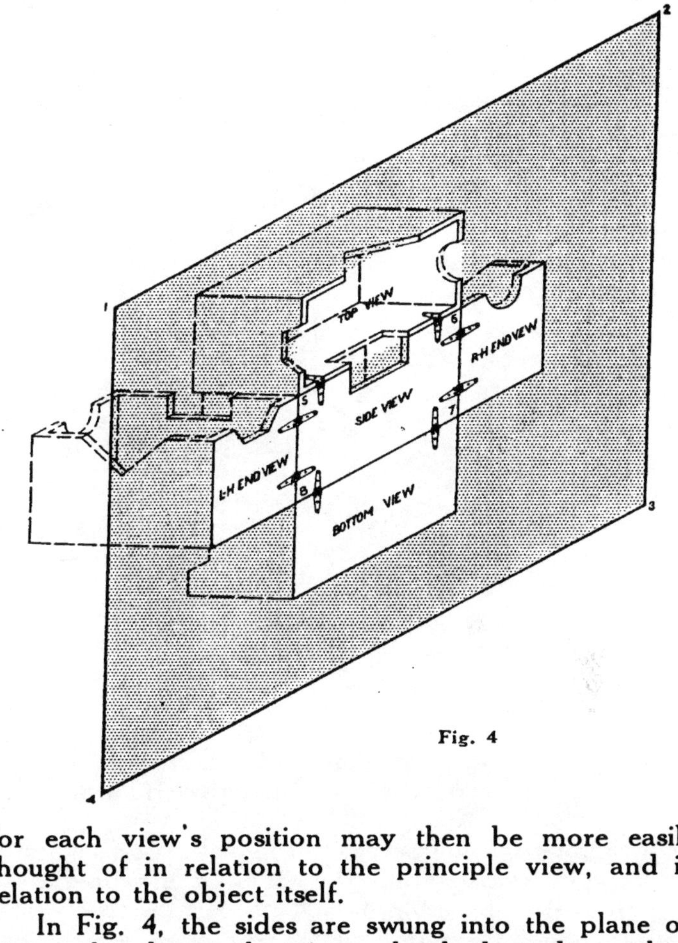

Fig. 4

for each view's position may then be more easily thought of in relation to the principle view, and in relation to the object itself.

In Fig. 4, the sides are swung into the plane on hinges that keeps the views closely butted together.

The relation between the views is clearly understood under these conditions. It is true, of course, that so far as relative position is concerned, the relation of the views, one to another is not changed, but only made clearer when the views are separated as in Fig. 5. You

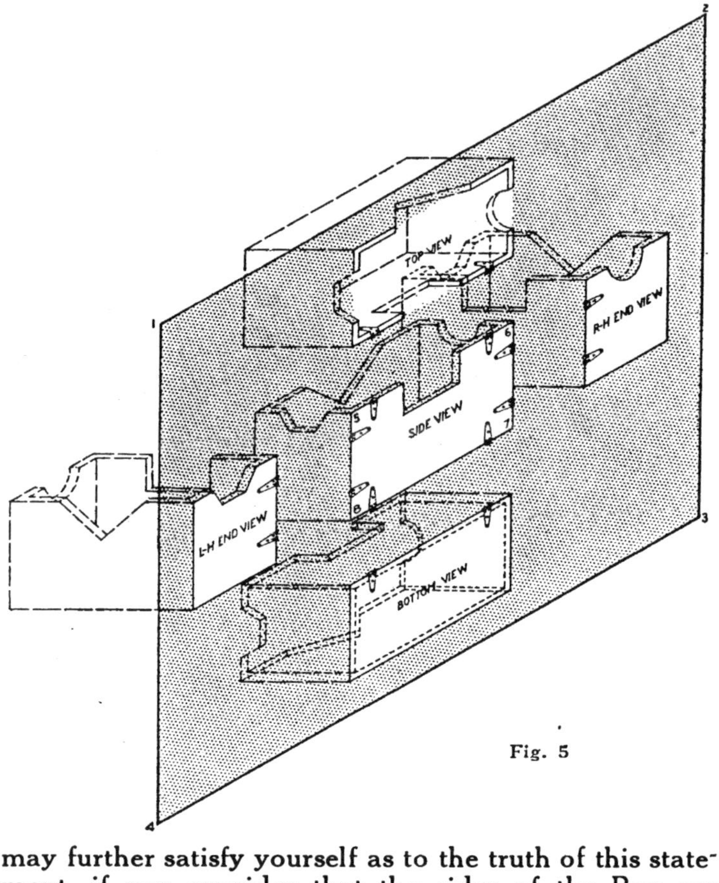

Fig. 5

may further satisfy yourself as to the truth of this statement, if you consider that the sides of the Box are swung up to the plane 1-2-3-4, in Fig. 4 in such a way that all the sides are butted together, making each side with the side that is next to it look as if all of the sides are in one view. Imagine that to change this crowding together, you remove all of the hinge pins from the hinges and separate each view of the Boxes shown in Fig. 5.

Having each view of the Box separated as in Fig. 5, the Box itself is shown by means of dashed lines in the position that is imagined to occupy when any one of the views is considered. Therefore, by means of dashed lines, the Box is shown in exact relation to each one of the views as it is considered.

The complete Box as represented in the several positions in Fig. 5 as extending in back of each view on the plane, will aid the student in understanding the relation of one side of the Box as a view on the plane to any other side of the Box as a view on the plane.

In Fig. 6, the Box is shown as in Fig. 5, only that the hinges are removed.

As each of these views on the plane, as in Fig. 6, is shown with a complete outline of the Box extended back of the plane, each view shows by the use of dotted (hidden) lines all that is back of the surface of the view. This requires the use of imaginary projection lines such as shown in Fig. 6. These projection lines are only imaginary. They project from points which lie straight back of each view as is shown drawn on the plane.

Occasionally, so called mechanics are seen, who in order to project the side of an object into a view on a plane, swing their hands in the direction that the imaginary hinge swings. While this practice is excusable in one who has had little experience at the reading of blue print drawings, to one who really knows how to read a blue print drawing, it conveys an impression of inability to cope with the principle of **projection** so easily mastered.

The projection lines shown in Fig. 6 will be more fully explained under the subject of "PROJECTION OF LINES AND POINTS." Before reaching that subject, it will be well to note in Fig. 6 that the lines and points that each view shows, are **projected** or brought up to the plane.

What has been explained regarding the use of full and dotted (hidden) lines applies in the same way to **projection** lines.

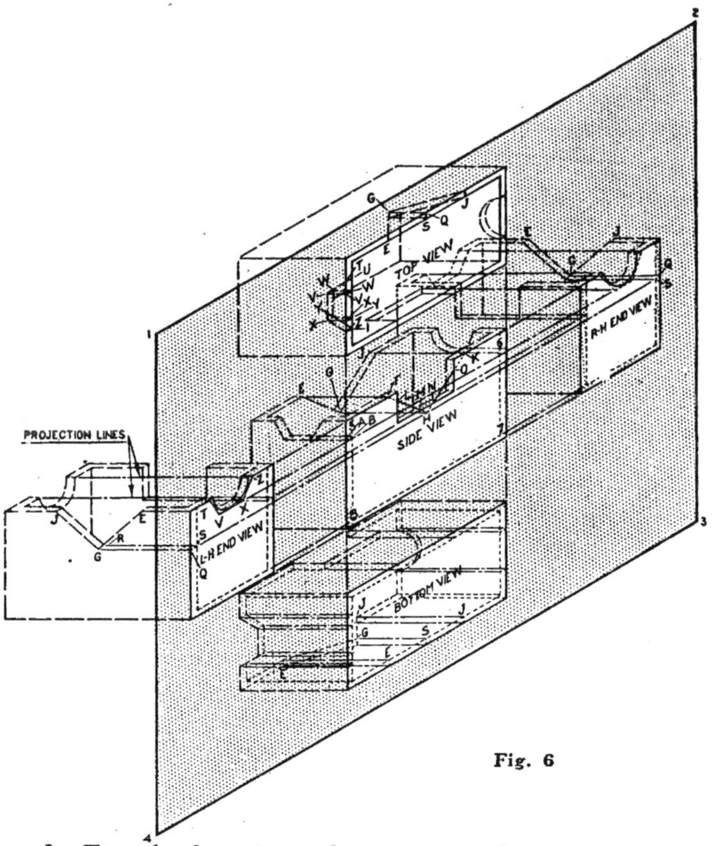

Fig. 6

In Fig. 6, the views that are openly seen are projected on the plane in full lines. The parts of the object in back of the plane hidden actually by surfaces nearer the plane are projected on the plane in dotted lines.

The dotted (hidden) lines on the plane in the bottom view of Fig. 6 represent lines which are not openly seen. They represent hidden lines showing the thickness of the sides of the Box, and may be thought of as occupying a place in back of the plane straight opposite the bottom view. The bottom of the Box is thus nearer the plane and hides the top of the Box from actual sight.

Of course, in the top view in Fig. 6, the position of the Box back of the plane is reversed. The top of the Box in this instance conceals the outside bottom.

The thickness of the sides of the Box in Fig. 6, it will be observed, cannot be openly seen, except in the top view, it is therefore shown in all the other views with dotted (hidden) lines. The student should of his own accord find the relation of hidden and open surfaces in each of the other views in Fig. 6, and note the use of full and dotted lines.

Figure 6 offers opportunity for an extended study of **line projections.** The student should take ample time to master the principles illustrated in this figure, especially as they apply to **projection lines,** as well as the use of full and dotted lines.

In Fig. 7 is shown the plane 1-2-3-4 with the sides of a Box **projected** as Views on a Plane. The method of obtaining these views has been explained in connection with Figs. 3-4-5 and 6.

It was with the purpose of securing the full meaning of blue print drawings that the foregoing illustrations of a complete Box extending back of each view with **dashed** lines was shown. The representation of the Box extending back of the plane in the various views, were intended to give the student a clear understanding of the relation between the whole Box and each of the views, and of the relation of each view to each other view.

All of the views shown in Fig. 7 were drawn exactly as were those shown in Fig. 4, and were separated as were those shown in Figs. 5 and 6. The view drawings in Fig. 7 are simpler than those in either Figs. 5 and 6, because in Fig. 7, the views are shown without any other lines to confuse you.

In all other blue print drawings, each view shown is to be thought of in relation to all other views exactly

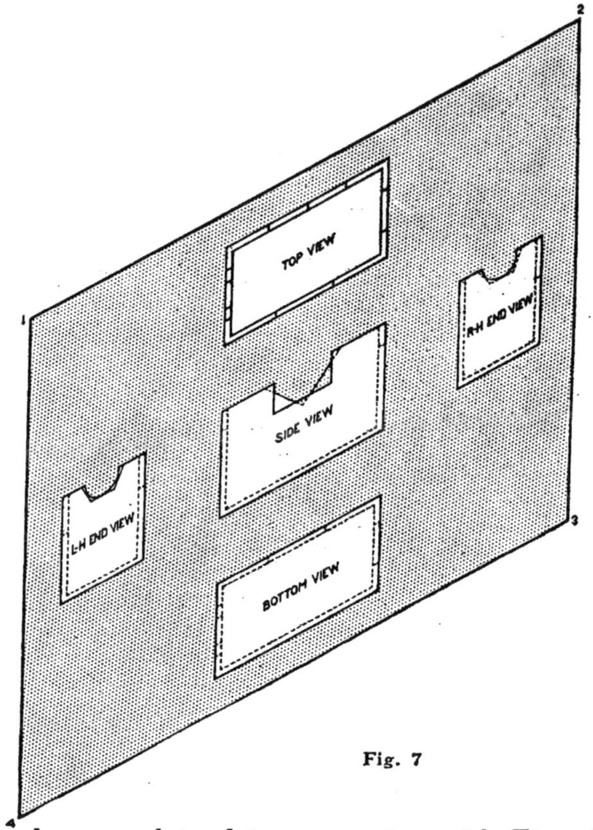

Fig. 7

as has been explained in connection with Figs. 3-4-5 and 6 that illustrates the views and **projections** of the Box.

The skillful blue print **reader** must learn to see in the blue print drawing more than is actually set down in lines. The views in Fig. 7, for example, should suggest to the reader the various relations shown by dotted lines in Figs. 5 and 6. Figure 7 with its several simple views should give as clear ideas of the Box and its parts as do the views and projections in Figs. 5 and 6. When the student begins to feel that this is so, he may then rest assured that he understands the fundamental principles of **straight view projection.** An understand-

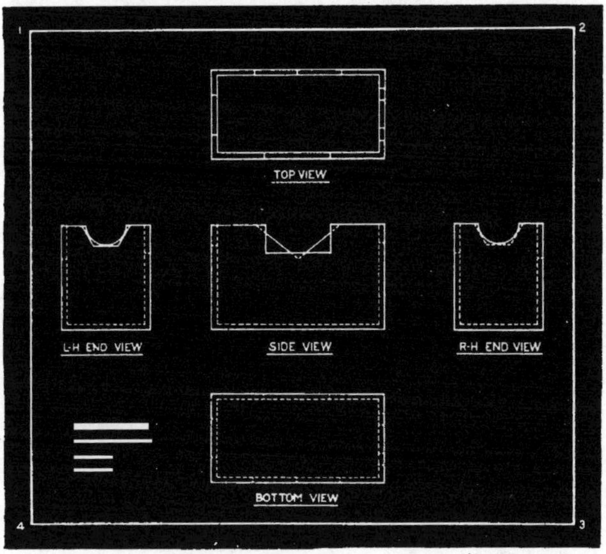

Fig. 8

ing of the above principles is the basis of mastery in the reading of blue print drawings.

In Fig. 8 the plane is shown lying straight, but in all other respects bears the same to the plane as in Fig. 7. Figure 8 shows every detail of a complete Box as it should be drawn, and as a blue print drawing should represent it under the conditions that have already been stated.

In Fig. 9 is illustrated a method of **projecting** lines from one view to another view, which should be of considerable help in finding where one line or point of one view can be located on another view, should these imagined **projection** lines be followed from view to view. The above method can be applied to the reading of any blue print drawing.

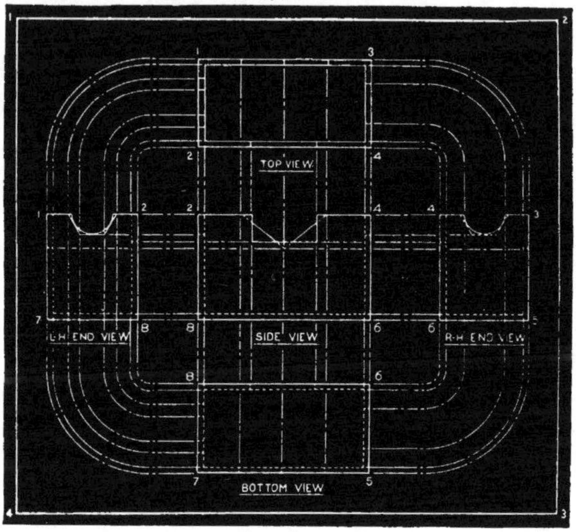

Fig. 9

There is a figure on each corner of the Box of Fig. 9. By following the figures that are alike, you can match the corners that are the same.

It would be well to re-read and follow the explanations given with the various cuts and figures given so far until you are very confident that you thoroughly understand the development from the perspective drawing of the Box in Fig. 1, to the complete blue print drawing of the Box shown in Fig. 8.

To test your understanding of the principles so far explained, you should read step by step only what the illustrations show in Figs. 1 to 8 and explain these illustrations in your own words.

ANOTHER METHOD OF STRAIGHT VIEW PROJECTION

Another method of obtaining **straight view projections** may be illustrated with the drawing of the Claw Hammer, commonly called the Carpenter's Hammer, which is a tool that every man has seen and handled.

In Fig. 10 is shown a perspective view of a Claw Hammer, in other words, a picture of a Claw Hammer as your eye generally sees it. A perspective drawing of Fig. 10 convevs no particular information as to the construction of the Claw Hammer.

You will note from letter "A" in the perspective drawing of Fig. 10, that the hole which holds the Handle suggests very little detail as to the hole's shape. From what is seen at the end of the Claw Hammer in Fig. 10, it is not known whether the hole is straight or tapered, or how wide this hole is. The proper shaoe of the Handle hole can only be had from what the separate views show in the blue print drawing of Fig. 11.

When a Hammer is represented, as in Fig. 11 blue print drawing, the dimensions and shape can be provided for each particular part. A perspective drawing, like that of Fig. 10 could not conveniently show dimensions.

The blue print drawing of Fig. 11 shows the correct shape for each part that is seen in a side of the Hammer as a view, besides the placing of dimensions can be arranged wherever needed. Of course, every blue print drawing of any other article provides likewise.

To understand clearly the method of obtaining the views shown in Fig. 11, is to imagine that the lines drawn about the perspective drawing of the Claw Hammer in Fig. 10, is to represent the corners of a glass box, and that your Hammer fits snugly into this box.

The corners of this glass box are numbered 1-2-3-4-5-6-7 and 8 for convenience in referring to each side as a view of the Hammer, as it is necessary to refer to these sides separately in making a clear explanation.

Since the Hammer fits snugly into this box of Fig. 10, you look into one, and only one of the sides of the box, the walls of this box should prevent your seeing any other side of the Hammer, except that seen straight in back of the side that you look through. Only one side of the Hammer is then clearly in view. Suppose, for example, that you look through the 2-4-6-8 side of the glass box, paying no attention to the other sides for the present. This would give you the side view shown in Fig. 11.

The side view of Fig. 11 gives only those details of shape as belongs strictly to the side of the Hammer seen through the side 2-4-6-8 of the box.

In the left hand end view of Fig. 11, there are shown principally the width and height. The length, along with the curving shape and general outline can be seen only in the side, top or bottom views. The position of the eyes in looking at the L-H. end view of Fig. 11 is like the position of the eyes in looking through the L-H. end view 1-2-7-8 of the box of Fig. 10.

The right hand end view of Fig. 11 is like the position of the eyes when looking at the R-H. end view through the side 3-4-5-6 of the box in Fig. 10.

When looking at any view shown in Fig. 11, think of the eyes in a position as illustrated in Fig. 10, by looking through a particular side of the box that shows the side of the Hammer represented in the particular view.

It is expected of you to look into the side 1-2-3-4 of the box in Fig. 10 while getting an understanding of the top view, and into the side 5-6-7-8 of the box to get an understanding of the bottom view, as shown in the blue print drawing of the Hammer in Fig. 11.

To understand more of the Hammer, as to how the Handle fits into the Hammer, is to look at the dotted (hidden) lines in the center of the side view of Fig. 11, shows not only the Handle fitting into the Hammer, but also the length and height of the hole into which the Handle fits. To see how wide the Hammer Handle hole is, you must look at either of the end views, or the top or bottom views of Fig. 11.

A part of the Handle can be seen in the top view, hence this part is shown with full lines. What cannot

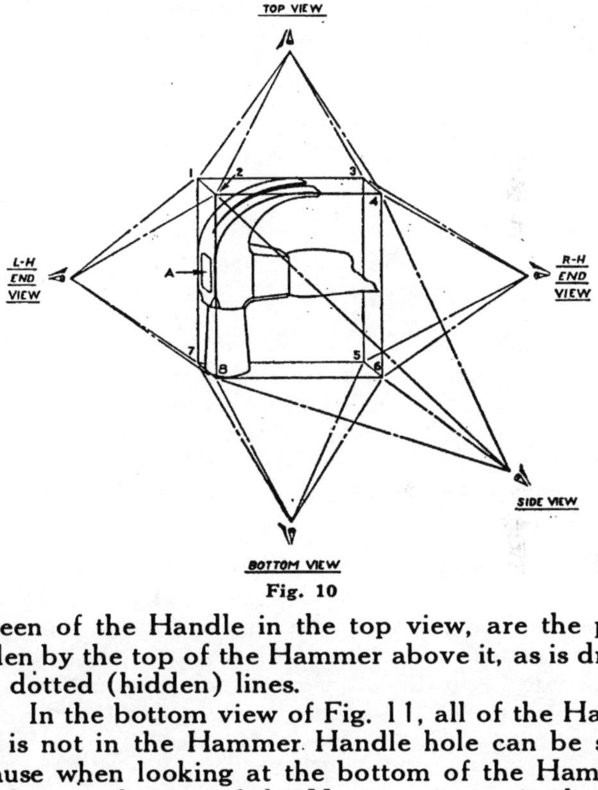

Fig. 10

be seen of the Handle in the top view, are the parts hidden by the top of the Hammer above it, as is drawn with dotted (hidden) lines.

In the bottom view of Fig. 11, all of the Handle that is not in the Hammer Handle hole can be seen, because when looking at the bottom of the Hammer, the claw on the top of the Hammer is not in the way of a full open view.

In the top view of Fig. 11, the dotted lines show that the top of the Hammer has a receding curve which spreads across and is wider than the points that are below this top surface. The lower portion of the Hammer is thus covered from actual sight, though the blue print drawing reveals all lines that can be **projected** into the plane on the view in question.

The small dotted line circle in the top view of Fig. 11 represents the shape shown for the "N"-"O" diameter part of the Hammer given as a short straight line in the side view.

The **projection** (broken) **lines** that are seen from the small dotted line circle in the top view of Fig. 11 are projected through to the side view until the points "N"-"O" are reached, which makes the points "N"-"O" in the side view, the very same points "n"-"o" as seen in the top view.

The points "n"-"o" as seen in the top view are exactly on the center line 15 and 20, and are the very same points "N"-"O" as those seen in the side view. The center line 15 and 20 of the top view is all of the plane that the side view occupied, because the side view is **projected** from the center line 15 and 20, to the side view plane.

The hitting surface of the bottom part of the Hammer in the bottom view of Fig. 11 is shown in a full line circle, because when looking at the Hammer at its bottom or underside, there is nothing to obstruct a full open view. On the other hand, only a portion of the hitting surface of the Hammer is seen in the top view. This portion of the hitting surface is drawn in full line small curves "L"-"K," "F"-"H" and "I"-"J". These short full line curves in the top view represent the portion not covered by the top part of the Hammer.

The other letters in the different views of Fig. 11 show where the points as placed by these letters in one view are to be found in another related view. You will note that the lines in one view are just the reverse in another view, with either dotted or full lines, as the case may be.

If you do not understand fully the explanations just given of the projection of views of the Claw Hammer, you should secure a Claw Hammer like the one in Fig. 10, and by looking squarely at each side of the Hammer, study it from the standpoint of the view drawings of Fig. 11 of that part.

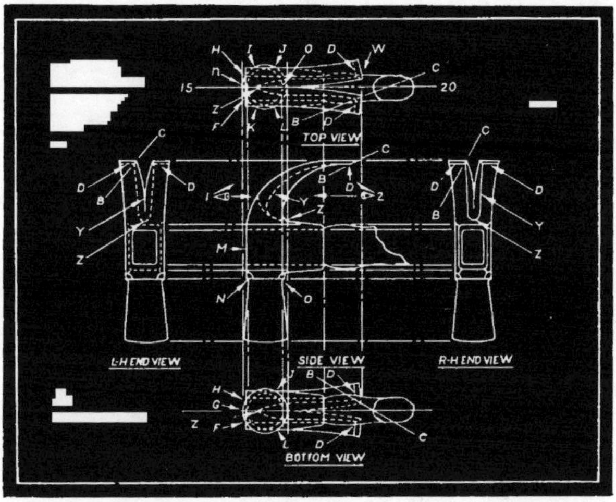

Fig. 11

By comparing the detail, the parts as seen in the side as a view of the Hammer, each view shown in the blue print drawing of Fig. 11 will be easily understood.

THE THIRD METHOD OF STRAIGHT VIEW PROJECTION

In Fig. 12 is shown a perspective view of a Machinist's "V" Block which is represented by the drawings of Figs. 13 and 14 illustrating another way to **project** the sides of an object as views on a plane.

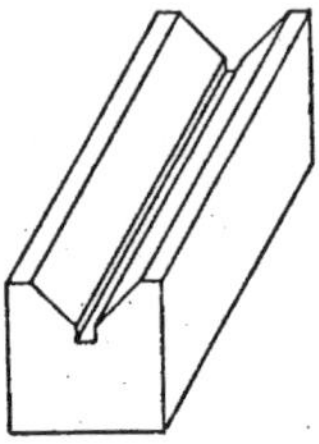

Fig. 12

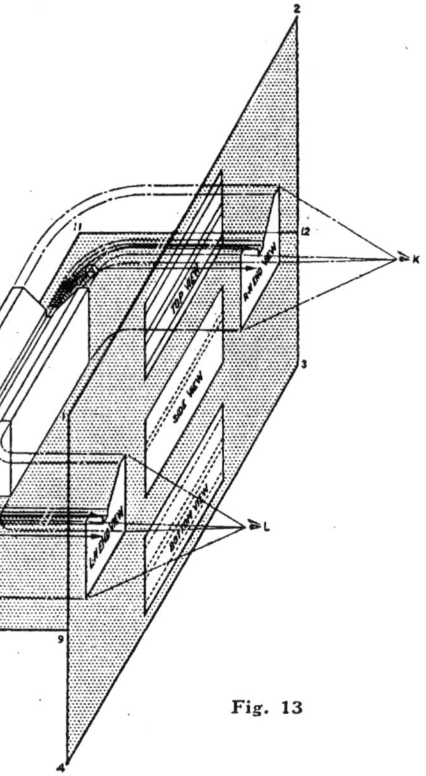

Fig. 13

Having the plane 9-10-11-12 as shown in Fig. 13 which is lying flat and square to the plane 1-2-3-4, and the "V" Block as shown in Fig. 12 placed on the end of the plane 9-10-11-12, one can represent by the use of **projection lines** each surface of the "V" Block as a view **projected** upon the plane 1-2-3-4.

In Fig. 13 is shown how the ends of the "V" Block are **projected** with **projection lines** to the plane for each

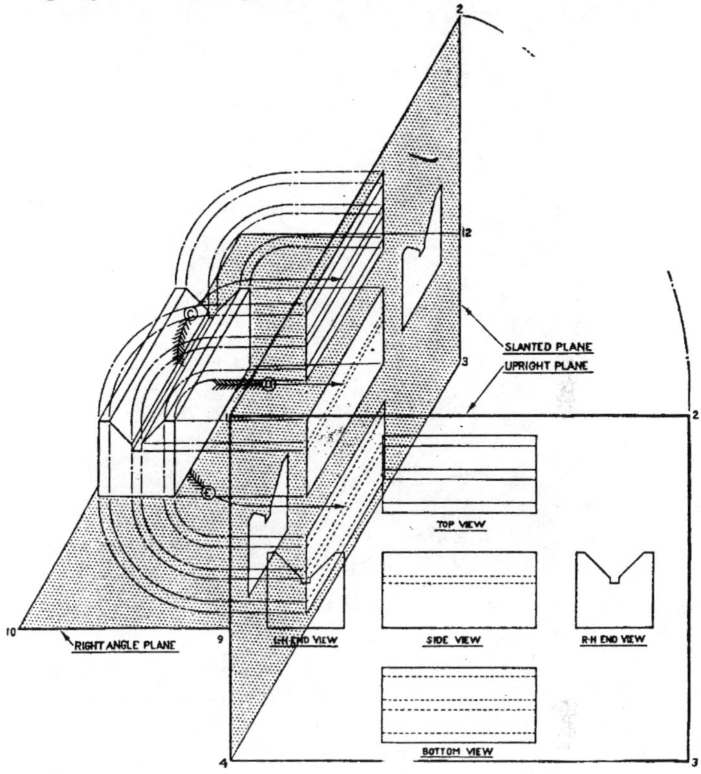

Fig. 14

of the end views. The R-H end view is shown by the course that the arrow "A" takes from the object to the plane, while the arrow "B" shows the course that the L-H end view takes to the plane.

To understand how the top, the bottom and the side views of the "V" Block are projected to the plane 1-2-3-4, the student should follow the course of the

projection lines in the direction of the arrows "C"-"D"-"E" direct from the object to the plane as shown in Fig. 14.

What is **projected** to the slanted plane 1-2-3-4 of the "V" Block in Figs. 13 and 14, is also **projected** to the upright plane 1-2-3-4. This upright plane 1-2-3-4 is the natural position in which to show every plane. From now on, in the explanation of how to read blue print drawings, the plane will be shown in this natural position.

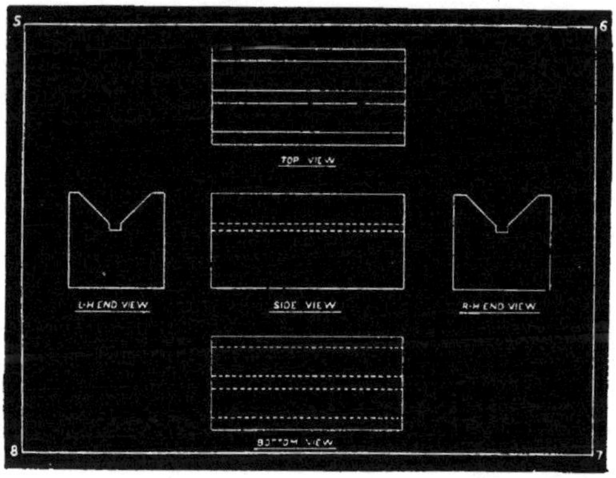

Fig. 15

The plane separated from the slanted representation of Fig. 14 is shown in its natural position in Fig. 15.

In Fig. 15, it will be clearly seen after a little study that all of the views are not needed to convey full information concerning the shape of the "V" Block, for each of the end views gives the same outline shape, and the same lines, hidden or open, may be shown in either the top view or the bottom view.

Either the top view or the bottom view, and either one of the end views are the only views needed to work a combination with the side view for a complete representation of the "V" Block blue print drawing.

Blue print drawings as previously explained, are a group of one or more views drawn to show the shape of one or more sides of an object represented by lines **projected** upon a plane.

So far, for the convenience of explanation, we have shown every side of the object projected upon the

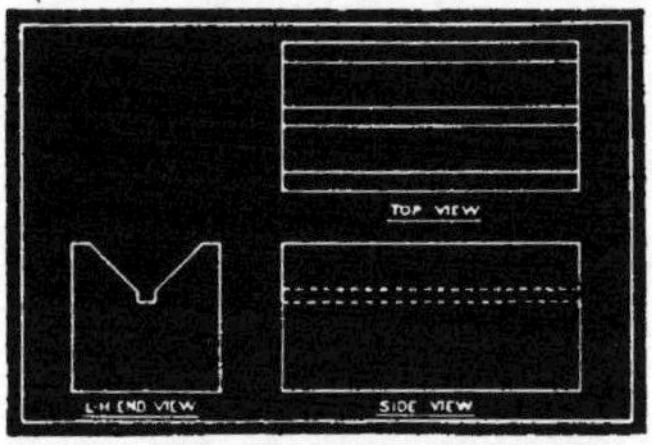

Fig. 16

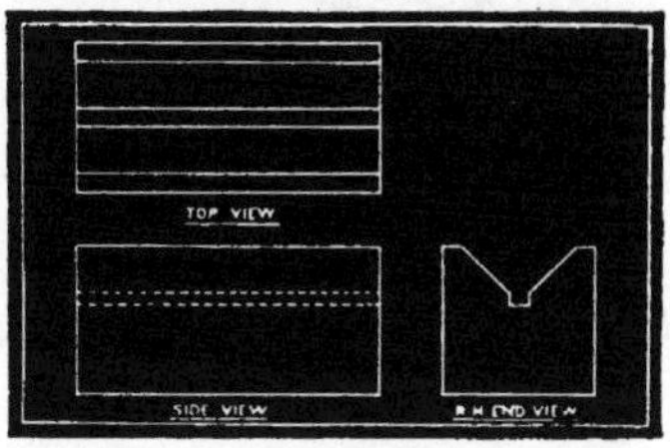

Fig. 17

plane, as in Figs. 8, 11 and 15, but every side of the object does not need to be shown in the blue print drawing, for in most cases, only three views are necessary. The three views most generally shown are the side, end and top views.

In Figs. 16, 17, 18 and 19 are shown four separate blue print drawings of the "V" Block of Fig. 12. Each

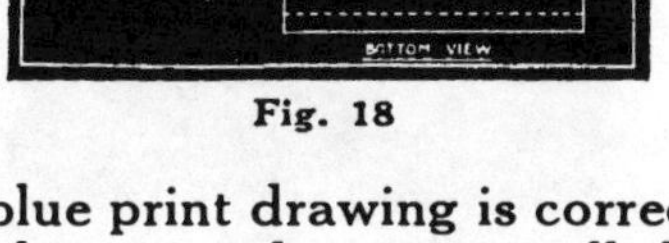

Fig. 18

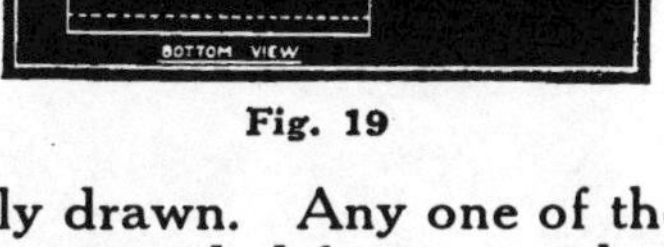

Fig. 19

blue print drawing is correctly drawn. Any one of the blue print drawings is all that is needed for a complete blue print drawing of the "V" Block.

Fig. 16 shows the side and the L-H end view with a top view. Figure 17 shows the side and the R-H end view with a top view. Figure 18 shows the side and the L-H. end view with a bottom view. While Fig. 19 shows the side and a R-H. end view with a bottom view. Each one of the above groups is drawn correctly, and any one could be used as a complete blue print drawing of the "V" Block of Fig. 12.

It is the practice of draftsmen to show in their drawings, views drawn with the least amount of dotted lines that is possible to show. You are not to infer from this that there should not be any dotted lines in any of the views, for the use of dotted lines depends upon what the views are to show.

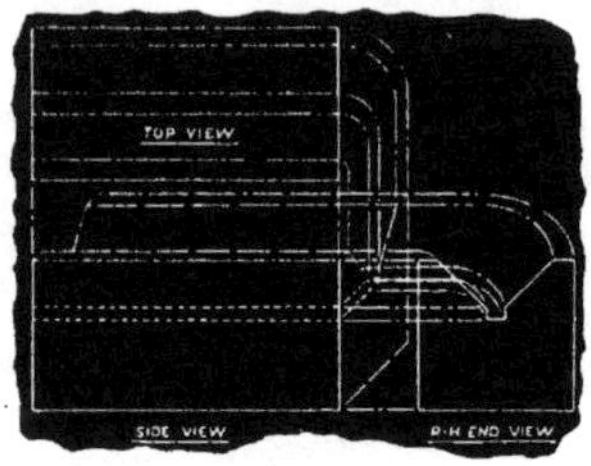

Fig. 20

The top views of Figs. 16 and 17 have no dotted lines, because there are no shapes below the top view to be shown in dotted (hidden) lines. The bottom view is undesirable, because that would have to be drawn with dotted lines.

Therefore, the views in either Fig. 16 or Fig. 17 would be the choice of a draftsman, although either of Figs. 18 and 19 might be drawn and be correctly understood.

As has been explained with reference to Figs. 16, 17, 18 and 19, that much depends upon the way a draftsman desires to arrange the views in a blue print drawing. He may use any one of the several arrangements to suit his purpose, yet there may be special advantages in a particular arrangement.

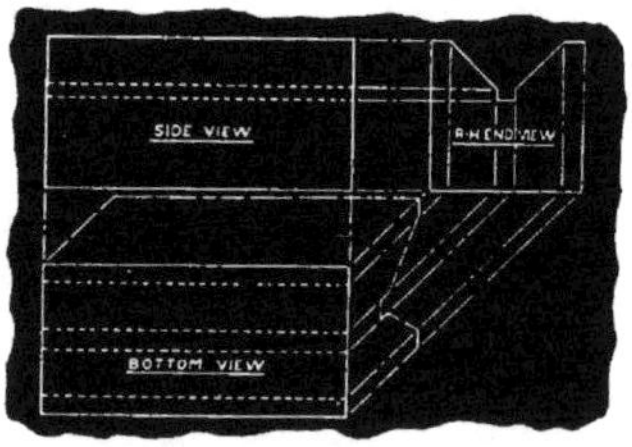

Fig. 21

A top view is always shown in preference to a bottom view, unless the bottom view is of such construction that it can not be shown drawn in the top view without too much complication of details, and only then is a bottom view generally used to avoid such undesirable complications.

What has been said regarding a choice between a top view or a bottom view, applies also to a choice between the two end views. Either one or the other of the end views is chosen, thereby avoiding the confusion that might easily arise through the use of too many views of the same class.

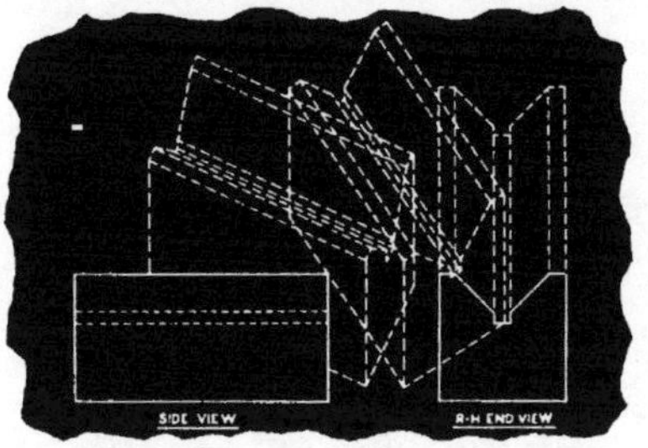

Fig. 22

As already stated, practically only three views are necessary for a blue print drawing of any article. It is well to note how Fig. 20 shows the **projection** of views for Fig. 17, likewise how the projection of Fig. 19 is to be thought of as illustrated in Fig. 21.

The **proper way** to project a view from another view is illustrated in Fig. 22, **and not** like Fig. 23 illustrates, which is the **wrong way.** Because of the danger of using this wrong method, much stress was given to the explanation of hinging of the Box in Figs. 3, 4, 5, 6 and 7.

It quite often happens that persons not thoroughly versed in the reading of blue print drawings undertake to project a L-H. end view upon the position of a R-H. end view as illustrated in Fig. 23.

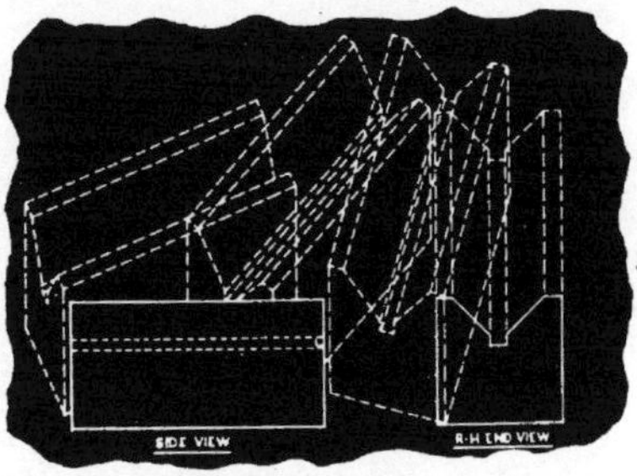

Fig. 23

It will be noted that the R-H. end view of Fig. 22 would be like its L-H. end view, because the object happens to be of simple construction with both ends alike. But, for example, consider the confusion that would result were there an attempt made to exchange the position of the end view of the Ratchet in the blue print drawings of Fig. 50.

PROJECTING TO A PLANE WHAT LIES BACK OF THE PLANE

Views **projected** upon a plane as explained so far must be thoroughly understood, that not only that part of the object in Fig. 24 that is lying actually in the plane that forms the view, but also the parts that lie back of the plane, must be **projected** upon the plane. These parts must be brought up to the plane from the dash lines that form the perspective object drawing of Fig. 24.

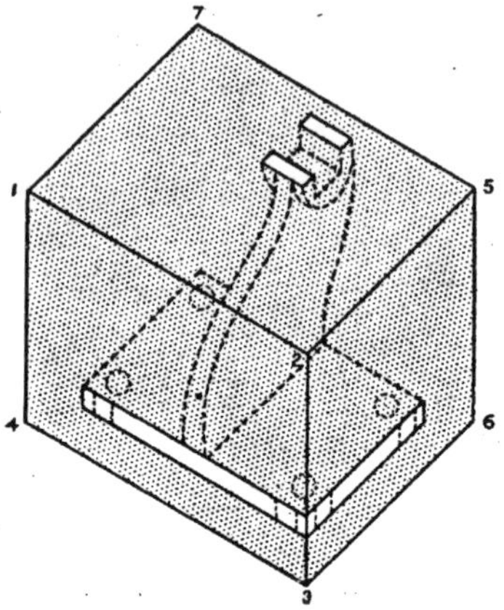

Fig. 24

Upon the plane 1-2-3-4 of Fig. 24 is shown only the edge of the base of the Bracket that lies in the plane, but this edge does not form a complete side view of the object. Everything extending back of the plane 1-2-3-4, must be brought up to the plane, so as to make the object a complete side view like that **projected** to the plane 1-2-3-4 in Fig. 25.

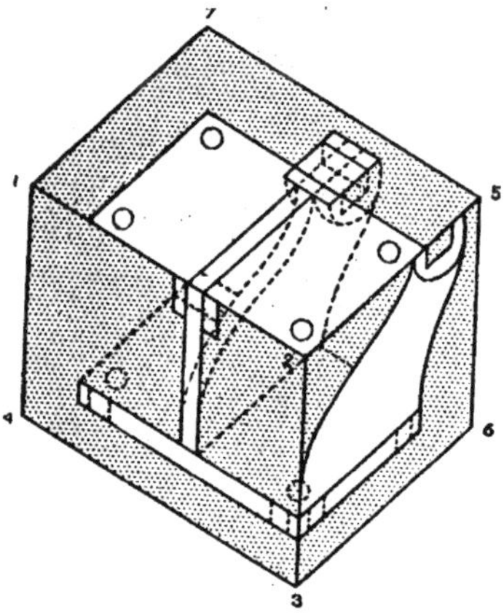

Fig. 25

What is projected on the plane 1-2-3-4 of Fig. 25 in full lines represents the outside shape of the object. Lines that cannot be actually seen on the plane 1-2-3-4, are the lines that are drawn dotted, because dotted lines in any kind of a view show hidden lines that cannot actually be seen.

Figure 26 illustrates a hinging motion to obtain views, as explained of the Box.

You will note from what has been said, that all full lines in a blue print drawing do not necessarily belong originally on the plane, because what is in back of the plane to be **projected** (brought up to) upon the plane, and represented in full lines, if it is in clear open view.

Figure 26 shows the three planes, 1-2-3-4, 7-5-2-1 and 2-5-6-3, which are no longer three planes as in Figs. 24 and 25, but opened out as one plane. These

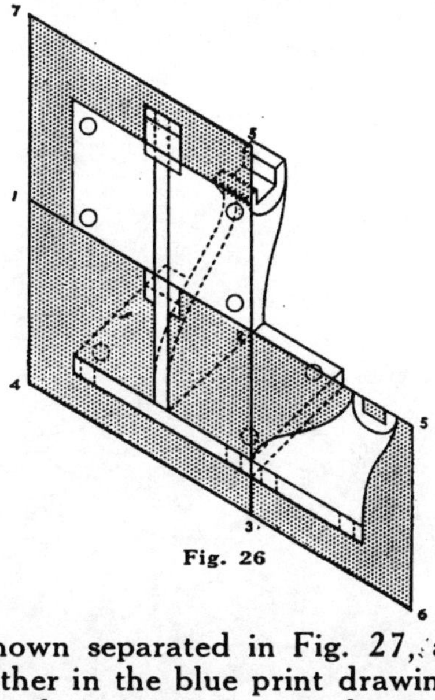

Fig. 26

planes are shown separated in Fig. 27, and grouped about each other in the blue print drawing of Fig. 28 in the same relative position as shown in Figs. 26 and 27.

By studying carefully the arrow lines in Fig. 28, and tracing the views from figure to figure, in Figs. 24 to 28, a clear understanding will be arrived at regarding the original position of the object represented, and the relative position of the views and methods of showing these views in the blue print drawings.

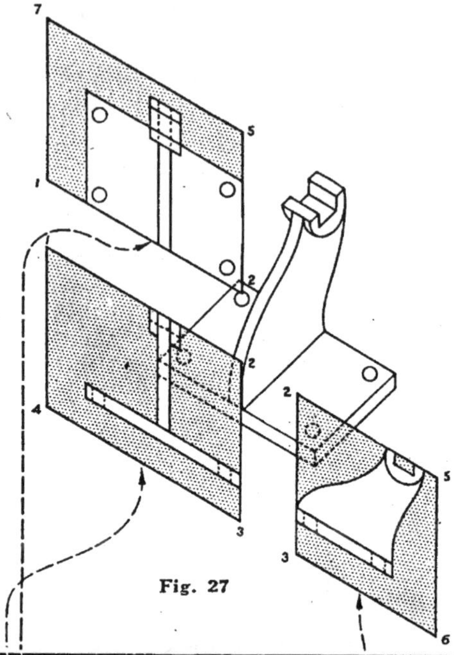

Fig. 27

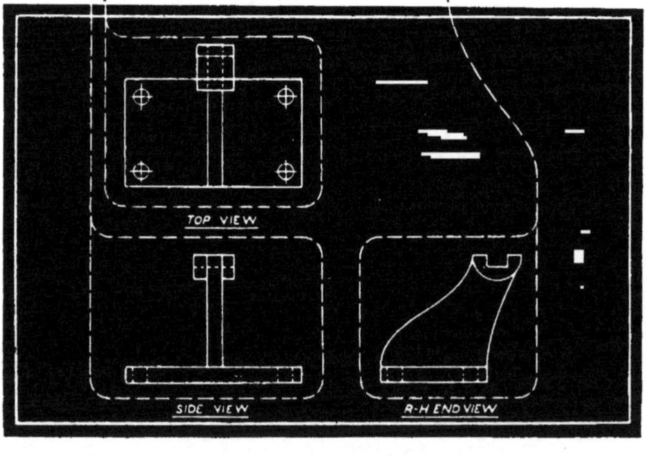

Fig. 28

PROJECTING OF LINES AND POINTS

Figure 30 is a complete blue print drawing of a Sliding Jaw of a Machinist's Vise, which are shown in four views. However, only three views, which can be the side and top, with either one of the end views, is all that is needed to give a complete blue print drawing.

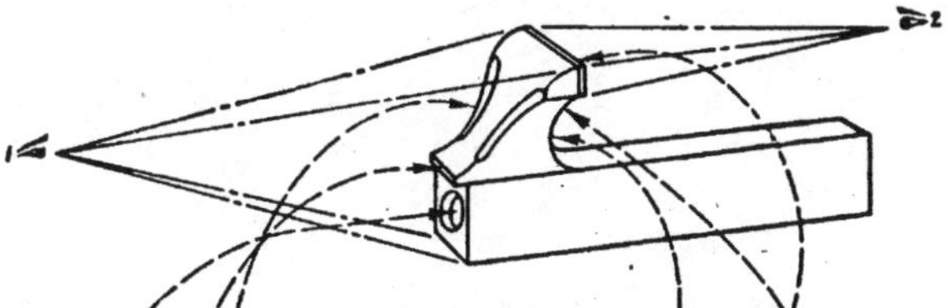

Fig. 29

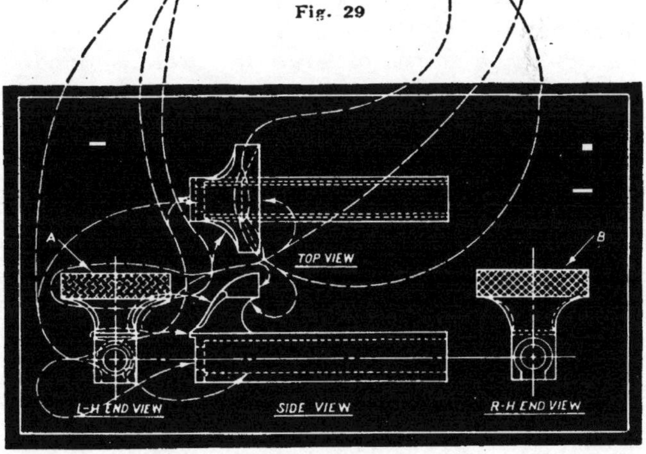

Fig. 30

The purpose in showing both the right and left hand end views of Fig. 30, is to show you why the face of the Jaw "A" as seen **projected** in the left hand end view from the side view **cannot be seen** in open view. For this reason, the face of the Jaw "A" is shown with dotted (hidden) lines, because the **projection is upon** the side view in the direction of eye position No. 1 as shown in Fig. 29, while the same Jaw "A," shown as "B" in the R-H. end view of Fig. 30 **can be seen,** because the Jaw "B" is open to full view in the R-H. end view, as it was **projected** from the side view in the direction of eye position No. 2, as imagined in the perspective drawing of Fig. 29.

In Fig. 31 is shown a perspective view of a Jaw Body of a Machinist's Vise which no doubt you are familiar with. There is good training in examining the actual perspective objects of Figs. 29 and 31 as are represented by blue print drawings, as the student learns thereby to picture objects in his mind's eye as he looks at the blue print drawing.

In Fig. 32 is shown a complete detail blue print drawing of the Jaw Body of the Vise as drawn in three views, as three views give all the details needed to show the entire construction of the Jaw Body part of the Vise.

By following the dashed arrow lines in Fig. 32 from one view of the object to its other views, and continuing the same dashed arrow lines to its perspective view of Fig. 31 you can readily appreciate how one **line** in a view, if **projected** to another view can be more fully understood. Each and every view of any blue print drawing contains the very same **lines** or **points,** but seen in their different view positions, and giving some additional source of information for each view.

When reading a blue print drawing, the reader should at all times train his mind upon all of the views, so that when **projecting** one or more **lines** from one view to that of another view, get in mind a picture, such as the perspective view of Fig. 31 shows for Fig. 32. Then such **lines** as make up the Jaw Body part shown in the blue print drawing of Fig. 32, will be as clearly seen and understood as the perspective Jaw Body view of the Vise in Fig. 31.

An understanding of the **projecting** of **lines** and

points is of great importance in the reading of blue print drawings.

The method explained in this chapter is practically the same in many respects as the method of **projecting** views to a plane, but instead of **projecting** a view to a

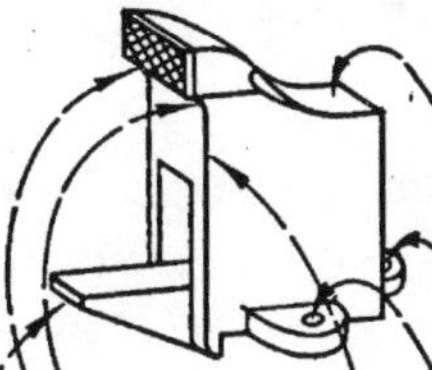

Fig. 31

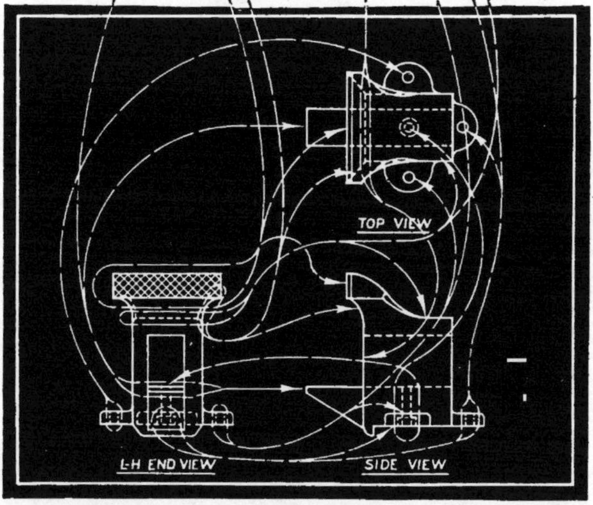

Fig. 32

plane, a **line** or **point** is **projected** from a part of one view to the views that are next to it.

The principle **projection lines** and **points** are shown extending from one view to another view in Fig. 32.

Figure 33 shows a Stand correctly represented in every respect in the blue print drawing in Fig. 34.

The lines connecting each view of Fig. 34 are **projecting lines** which are used, as has been explained, in tracing the position of **lines** or **points** from one view to another view.

Any particular **line** or **point** in a view of any blue print drawing that does not readily show the location of the **line** or **point** in its next view, must be looked for in the next view along a parallel to the edge **projection line** to that view, as shown in Fig. 34.

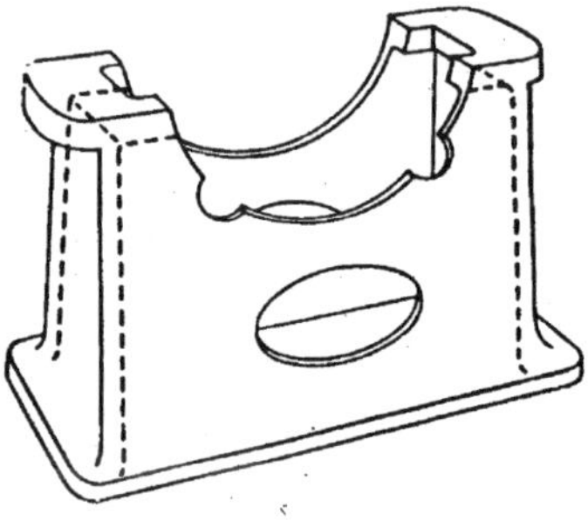

Fig. 33

In every blue print drawing, there is one view that suggests practically the main idea of the object drawn. Since this one view of a group of views offers so much, that view must be sought for the main source of information.

For example, consider the half round circles "W" and "X" of the side view of Fig. 34, which side view is the **main** view for understanding the whole blue print drawing, so far as the half round circles "W" and "X" are concerned in the various views of Fig. 34. To locate the half round circles "W" and "X" in the end view, look straight across the line "A"-"B," "C"-"D" and "E"-"F" from the **points** "A"-"C"-"E" of the side view, and as the **points** "A"-"C"-"E" of the side view are seen only on one side of the end view, then the half round circles "W"-"X" as seen in the side view are only to be on one side of the end view, which is the left hand side of the R-H. end view.

To confirm that the half round circles "W" and "X" as seen in the side view of Fig. 34 is shown only on one side of the end view, look at the top view of Fig. 34 along the straight upright **projection lines** as you did in tracing the half round circle **points** "W" and "X" in the end view.

As the **points** "A"-"C"-"E" of the side view are the only **points** that the upright **projection** lines of the half round circles "W" and "X" show on one side of

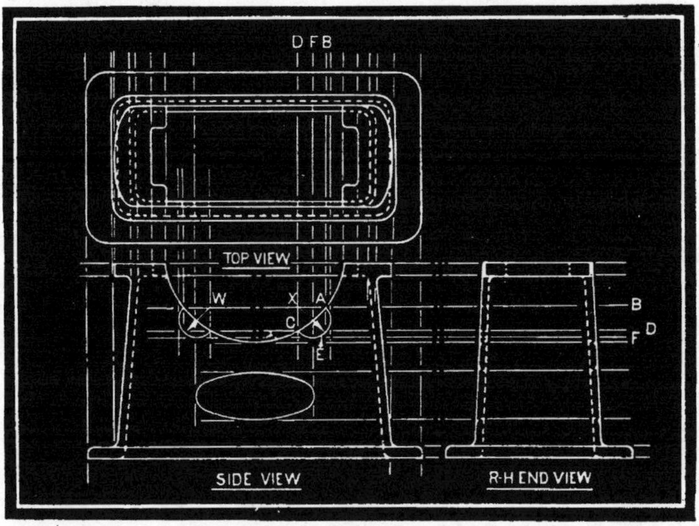

Fig. 34

the top view, it is known, therefore, that the half round circle openings "W" and "X" as shown of the side view of Fig. 34, are only on one side of the object, as represented in Fig. 33.

The two half round circles "W" and "X" as seen in the side view are each openly seen located in their respective positions in the top view of Fig. 34, because they are there in full open view, but the half round circle "W" and "X" are not both seen together in the end view as they are in the top view, because the half round circle "X" is first seen with dotted (hidden) lines in the end view, and the half round circle "W" is to be thought of as being straight in back of, and in line, and as far apart with the half round circle "X" as shown on the side view.

The lines showing the half round circle "X" **projected** from the side view of Fig. 34, to the end view, are also the same lines that show where the half round **circle** "W" is seen in the end view. The half round

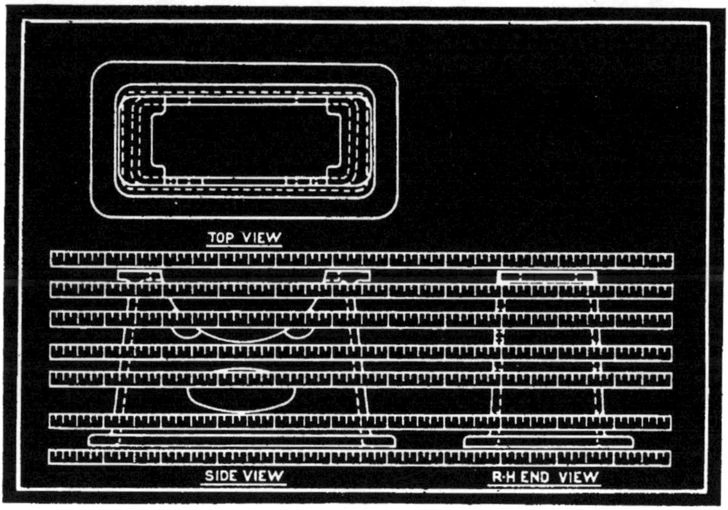

Fig. 35

circle "W" must be straight in back of the end view of the same distance apart as in the side view of Fig. 34 show.

The **projection lines** as shown in the blue print drawing of Fig. 34 to connect each view, are for the purpose of aiding the eyes to follow the lines to the same **point** on the other views.

Draftsmen should not draw these **projection lines** as in Fig. 34 on any kind of a blue print drawing, because the **projecting lines** would become confused with the lines of the drawing. The reader of a blue print drawing should imagine that he sees these **projection lines.** Draftsmen who draw such lines thereby show their inexperience.

Since blue print drawings should have no **projection lines** to connect the views as in Figs. 35 and 36,

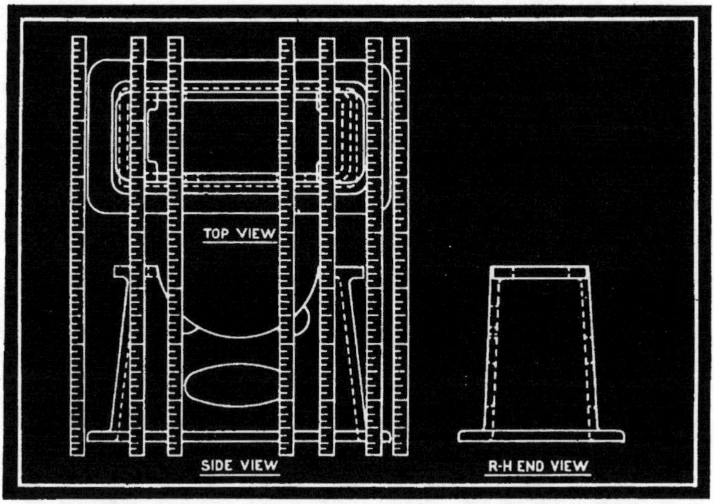

Fig. 36

it is a good practice for the beginner to use a straight edge or rule parallel to the edges of the blue print drawing as shown in Figs. 35 and 36.

It is well to note in Figs. 35 and 36 that each rule or straight edge serves as a guide in tracing a **line** or **point** of one view to the very same positions in the other views.

Fig. 37

Figure 37 shows a perspective drawing of the Ratchet disc as shown in the blue print drawing of Fig. 38.

The **projection lines** that connect the side with the end views of Fig. 38, show how the dotted as well as the full lines are used in representing the teeth in the end view. The lines of the outside edge of the teeth are shown with full lines in

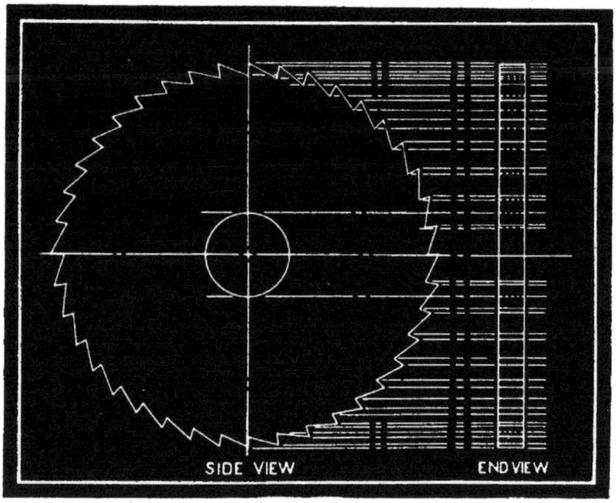

Fig. 38

the end view, because the outside edge is open to full view, while the dotted lines in the end view show the lower edges of the teeth, which are hidden by the upper edges, when **projected** from the side view.

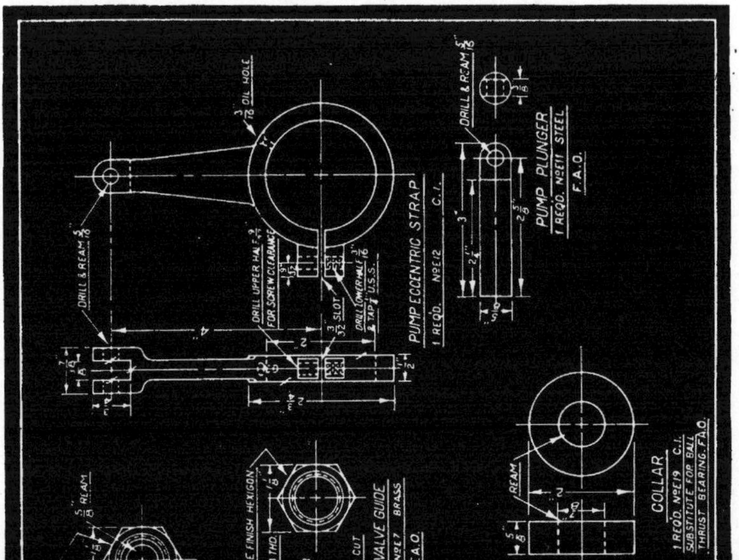

Practice Reading Figure

Figure 40 shows a complete blue print drawing of the four hole Block of Fig. 39.

In the top view of the blue print drawing of Fig. 40, are shown four holes lettered "A," "B," "C" and "D". These holes as shown in the top view of Fig. 40, must also be located in the side and end views, as shown with dotted (hidden) lines as in Fig. 39.

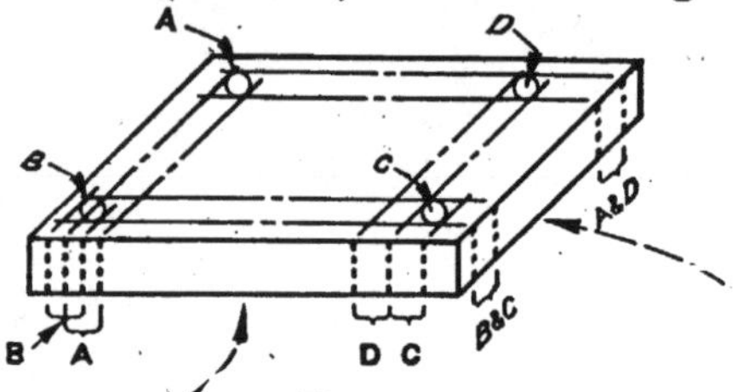

Fig. 39

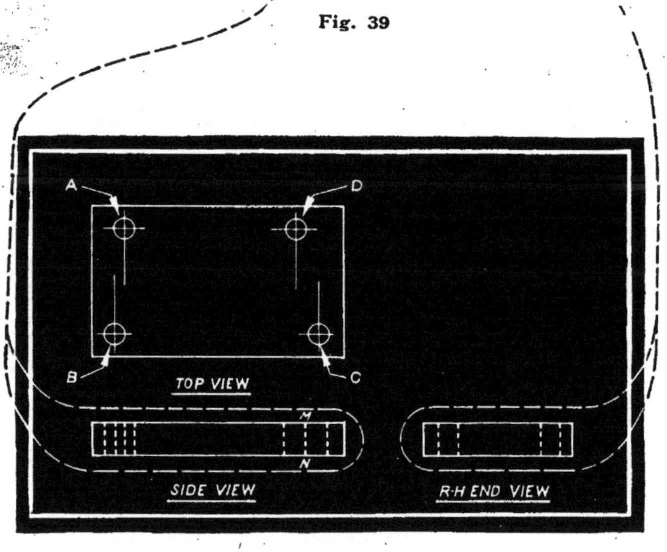

Fig. 40

The dotted (hidden) lines in the side view of Fig. 40, represent what is seen in the side view of the perspective drawing of Fig. 39. The dotted (hidden) lines for the sides of the holes "A" and "B" are intermingled in the side view as **projected** from the top view of Fig. 40.

The holes "C" and "D" of the right hand side of the top view of Fig. 40 are **projected** to its side view. One side of each of the holes "C" and "D" of the side view intermingle and show only the line "M"-"N," because each hole as seen in the top view is offset from the other just far enough, so that the intermingling of the dotted line "M"-"N" in the side view seems to be only one line, although it really represents two lines, this is because the dotted line "M"-"N" of the side view represents the side of the holes "C" and "D" that are in line with each other.

In the end view of Fig. 40 are dotted lines to represent four holes, but from first glance at the end view, only two holes seem to show.

The last paragraph should be borne in mind when looking at any view of any blue print drawing, that it is necessary to look at more than one view before deciding upon all facts concerning the object represented.

In the left hand side of the R-H. end view of Fig. 40, is seen the hole "C" **projection,** but straight in back of, and in line with the hole "C," is the hole "B," as is seen in the top view, which hole "B" is understood to be back of the hole "C," as both the holes "C" and "B" are apart in the end and top views.

The hole "D" is also first seen in the right hand side of the R-H. end view of Fig. 40, but the hole "A" is straight in back of and in line with the hole "D."

The above paragraphs will be readily understood by the student if he carefully examines Fig. 39 in relation to Fig. 40.

EXAMPLES OF VIEW PROJECTIONS

Figure 41 shows a perspective view of the Bracket Slide represented by the blue print drawing of Fig. 42.

Each view of Fig. 42 is encased with a dashed arrow line pointing to the perspective view of Fig. 41, of where each view is obtained from the object, and

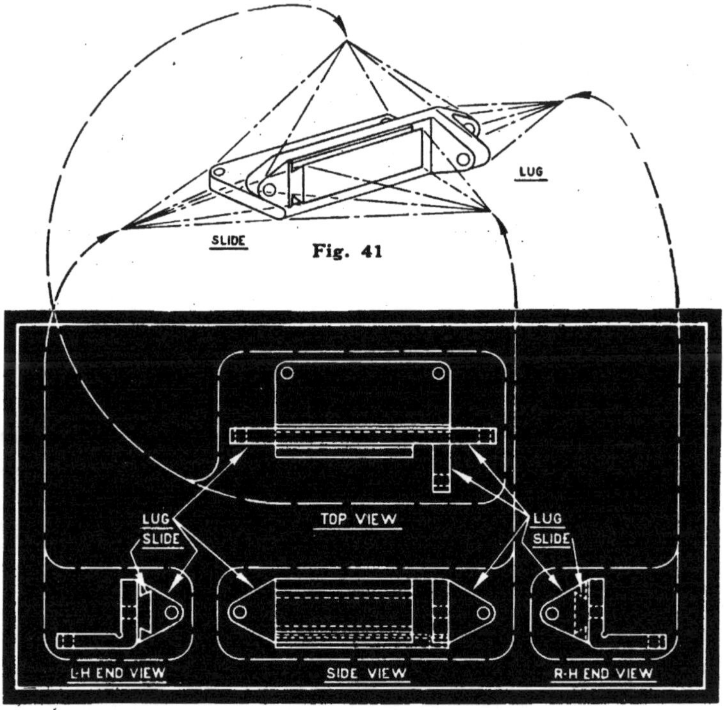

Fig. 41

Fig. 42

the relative position that each view in Fig. 42 has to each other view.

The four views as shown in Fig. 42 are more views than are needed. Either one of the end views of Fig. 42 is all that is necessary with the top and side views.

A complete blue print drawing of the object represented in Fig. 41 requires only a top, side and end view. Each of the end views of Fig. 42 conveys the same information, hence the use of either one of the end views will do to make a complete blue print drawing. The R-H end view will be used, as R-H end views are generally used in blue print drawings.

The R-H end view of Fig. 42 shows the Slide drawn in dotted (hidden) lines, because the Slide **is not openly seen,** as the Lug is in front of it. In the L-H end view, the Slide **can be openly seen,** hence the Slide is drawn in full lines, as that view has the Lug to the rear.

The encased dashed arrow lines around each end view of Fig. 42, and pointing to its side of Fig. 41, should give a good understanding why the Lug was drawn dotted in one end view and not in another.

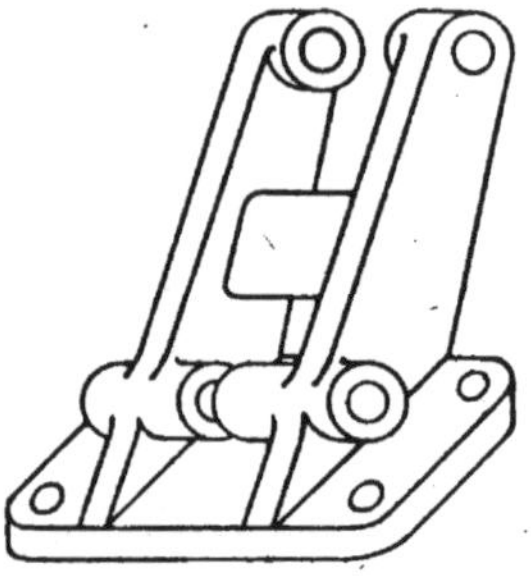

Fig. 43

Figure 43 shows a perspective view of an Arm Bracket which is fully represented in the blue print drawing of Fig. 44.

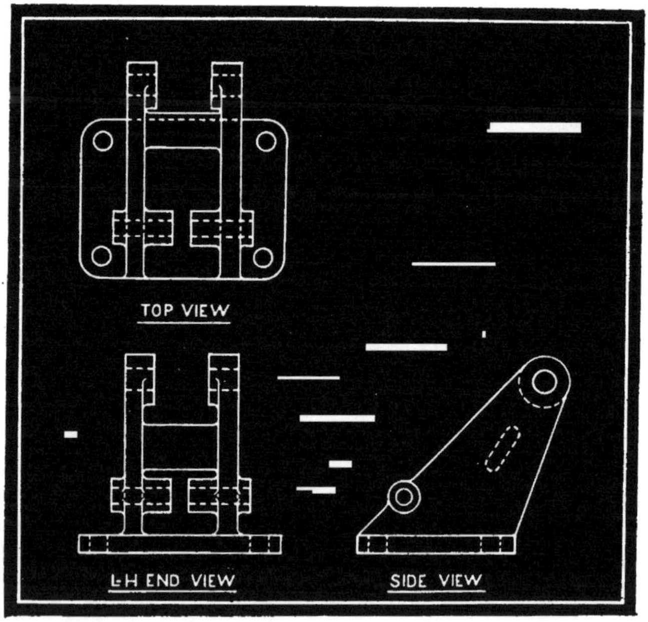

Fig. 44

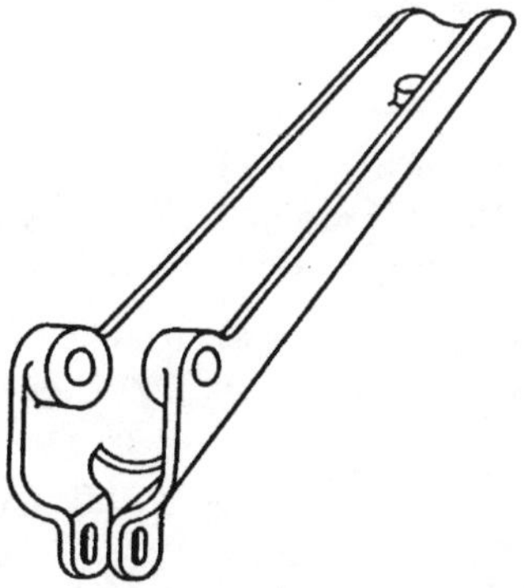

Fig. 45

Figure 45 shows a perspective view of a Lever Latch which is also fully represented in the blue print drawing of Fig. 46.

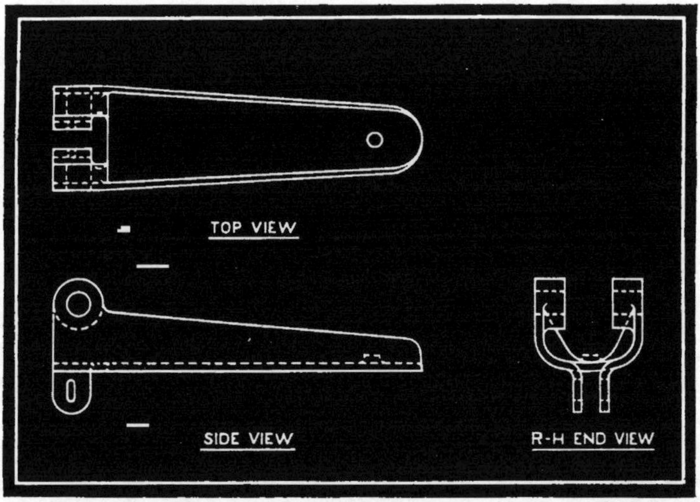

Fig. 46

The views shown in Figs. 41 to 46 should be used in reviewing the principles governing the **projection** of **views, lines** and **points.**

INFORMATION THAT ONE VIEW OFFERS ANOTHER

In Fig. 48 is show a blue print drawing of a Ratchet Wheel. The perspective drawing of Fig. 47 shows what the blue print drawing of Fig. 48 represents.

The blue print drawing of Fig. 48, has a side, a R-H end and a top view, which is enough information of the Ratchet shown in Fig. 47.

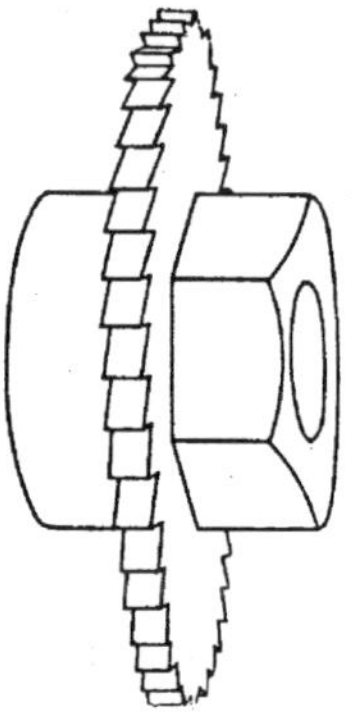

Fig. 47

Figure 48 could also show a L-H end view, but such a L-H. end view is not needed, as it would not add any additional information, for the side, end and top views as shown in the blue print drawing of Fig. 48, contain every detail that could be shown in a L-H. end view.

The object of this lesson on the Ratchet of Fig. 47, is to point out the importance of each view in relation to its other views of a blue print drawing.

To begin with, it is seen in the side view of Fig. 48, that the hexigon shape is seen drawn with dotted lines. As dotted lines always show what is hidden, then the hexigon as drawn in the side view must represent the back part of the Ratchet for that view. In the top and end views of Fig. 48, this hexigon shape can be seen openly.

Should the **"hinging method"** of **projecting** views to a plane as explained of the Box be applied to Fig. 48, it would show clearly and simply the side of each view on which the hexigon shape is found.

Truly, the blue print drawing of Fig. 48 is drawn in a simple manner so as to be clearly understood, as all blue print drawings should be. Blue print drawings are often seen, however, as shown in Fig. 50, which can be easily understood by one who thoroughly knows how to apply the fundamental principles, but at best, the blue print drawing of Fig. 50 would be puzzling and a stumbling block to most mechanics.

The cross sectioned end view of Fig. 50 has not the advantages of the views in Fig. 48, because in Fig. 50 there is not the hexigon and dotted tapered hole as that shown in the top and R-H. end views of Fig. 48 to guide you.

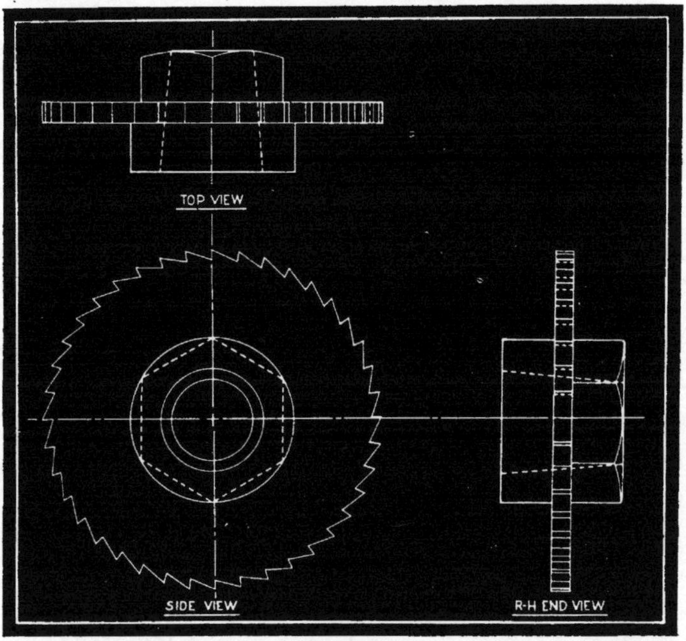

Fig. 48

It was intended by the draftsman to give no other consideration to an understanding of the cross sectioned end view as shown of the hexigon, other than that shown in the side view of Fig. 50, for the draftsman took it for granted, that those who were to use the blue print drawing of Fig. 50, would know how to read it.

The cross sectioned end view as shown in Fig. 50, is at this time out of the regular order of our explanation, as the subject of cross sections will be more fully explained under Part VI on "CROSS SECTIONS" to follow later.

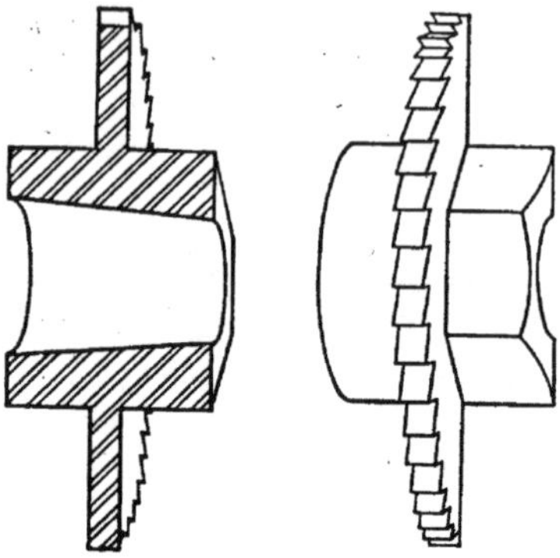

Fig. 49

Such blue print drawings as that shown in Fig. 50 reveal the real use of the **"hinging of views"** to a plane, that every one should know well enough to apply.

When looking at the end view of Fig. 50, the side of the Ratchet that the hexigon shape is on, may be easily found if the end view is **projected** from the side view to the plane in the same manner as explained of the Box.

To find which side of the end view of Fig. 50 that the hexigon shape is on, is to note that the hexigon shape in the side view is dotted, which means that the hexigon shape is hidden from full view, and that there must be something in front of it. Well then, the circle "A" of Fig. 50 is drawn with full lines, hence circle "A" **must be** in full open view, and on the front side of the side view. If circle "A" is in front, then the hexigon **must be** back of the side view. By **projecting** the views to a plane, as explained of the Box, it is an easy matter to find on what side of the end view of Fig. 50 that the hexigon shape is.

To strengthen the explanations given thus far on this subject, observe that the circle "A" which is seen in open view in the side view of Fig. 50 goes thru the point "B" where the hidden point of the hexigon meets the center line "1"-"2" of the side view, making with the circle "A," at the point "B," one common point. When this "B" point is **projected** across to the end view of Fig. 50, the straight line "b"-"b" is formed. Likewise the common point "C" when **projected,** forms the straight line "c"-"c," showing that the diameter for the round side of the Ratchet's end in the side view is the same as the extreme diameter of the points of the hexigon on the center line "1"-"2."

A mechanic who does not throughly know, after a little study, the meaning of such an end view as shown in Fig. 50, is not thoroughly versed in the principles of **projecting** a view to a plane, which he should

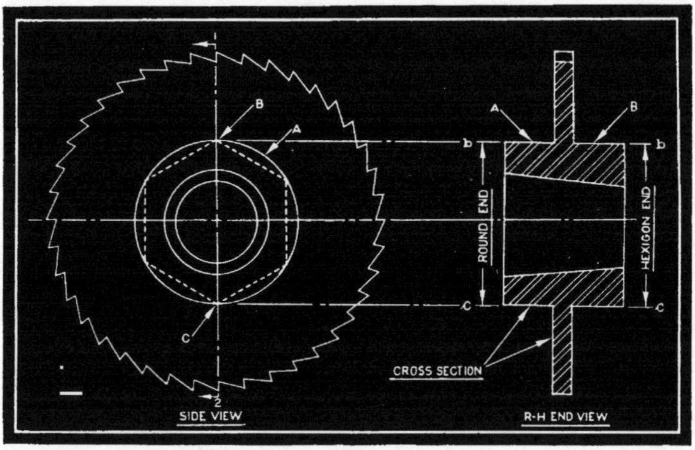

Fig. 50

above all things know thoroughly. Such an end view of Fig. 50 may ordinarily be of a more puzzling nature, as the comments "ROUND END" and "HEXIGON END" were only given on Fig. 50 to help aid in the explanation.

A good rule always to remember when seeking to get from one view, information of another view, is that in **projecting** a view to a plane, that the nearest side of the next view to the one you are considering, is the same side of the view that you are considering.

What has been explained of the Ratchet and its views of Fig. 50, applies in like manner to any other views in any other blue print drawing.

PART III

THE DISTINCTION BETWEEN VIEWS AND ELEVATIONS OR PLANS

The word "**view**" is used in naming the sides of an article when **projected** to a plane in a blue print drawing, as was used so far in these explanations.

The word "**view**" is not in the strict sense the proper word to use, although it is in general use among those who read blue print drawings.

The word "**view**" actually means what is seen of an object from any position that is within scope of the eye, which would apply more correctly to the perspective views as given in Figs. 1, 10, 12, etc. Therefore,

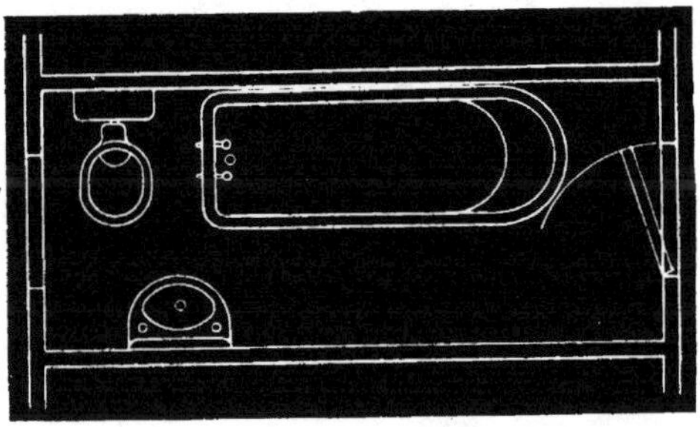

Fig. 51

such sides of an article as are **projected** to a plane should not in the proper sense be called "**views**," as that word conveys a hazy meaning. Instead of the word "**view**," use the word "**elevation**" or "**plan.**"

The word "**elevation**" in blue print drawings is used when speaking of the side of an object represented in an elevated or upright position, as a fence, or the side of a building, etc., showing no vanishing points, as that of a perspective view.

The word "**plan**" must not be confused with the word "**plane,**" for a plan is the representation of an article seen on a flat surface, such as the position a building occupies of the ground as may be seen from an airplane.

A good illustration of a **plan** is given in Fig. 51, showing the arraignment of the fixtures and the shape of a bath room floor. Another good example is seen in Fig. 52 showing the layout in the form of a map. The map is a good example of a **plan.**

In our explanations from now on, the side of any article as **projected** to a plane, which was previously

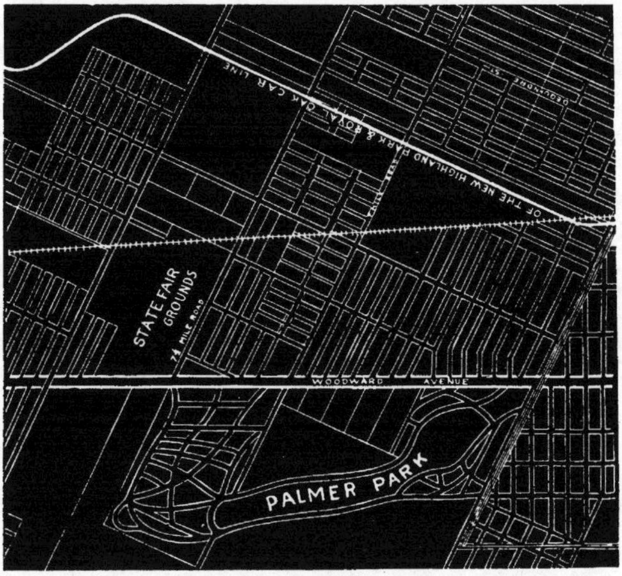

Fig. 52

called a view, is to be called an **elevation** or a **plan.** The following statements make clear the meaning of the words that will be used:

A side view is a **side elevation.**

A R-H. end view is a **R-H. end elevation.**

A L-H end view is a **L-H end elevation** likewise.

A top view is a **top plan,** and

A bottom view is a **bottom plan.**

PART IV

TWO ELEVATIONS OF THE SAME CLASS IN A BLUE PRINT DRAWING

A back elevation in a blue print drawing is a rare thing, for a back elevation is most generally combined into the front elevation by the use of dotted (hidden) lines. Should a back elevation be too complicated to give the proper details when drawn with dotted lines into the front elevation, then a back elevation is made into the drawing with a notation placed below, stating that it is such an elevation.

The blue print drawings of Fig. 55 is made up with sections "A" and "B," for the purpose of showing two elevations of the same class in a blue print drawing, instead of only one elevation of the same class as is usually done.

In showing the two side elevations, as the front and back (side) elevations of Fig. 55, the need of the two elevations of the same class for a blue print drawing is illustrated. In some blue print drawings there is too much detail to show in only one side elevation without an extra amount of dotted (hidden) lines that would be confusing. This is true of the blue print drawing in Fig. 55, which represents not only the front and back parts of the Slide Bracket, but also the slot "S" at the center.

The blue print drawing of section "A" of Fig. 55, is sufficient to show all that would ordinarily be required for Figs. 53 and 54.

In the blue print drawing of the Slide Bracket, in which there is the slot "S" at the center to be shown, as well as the front and back parts, two separate elevations of the same class are shown to make clear what the dotted lines represent of the slots.

The only reason for ever showing two side elevations as in the section "B" of the blue print drawing of Fig. 55, is to show the front (side) elevation in full lines (is open to full view), and the back (side) elevation in full lines (also open to full view), and to secure the difference of between lines not transposed from full

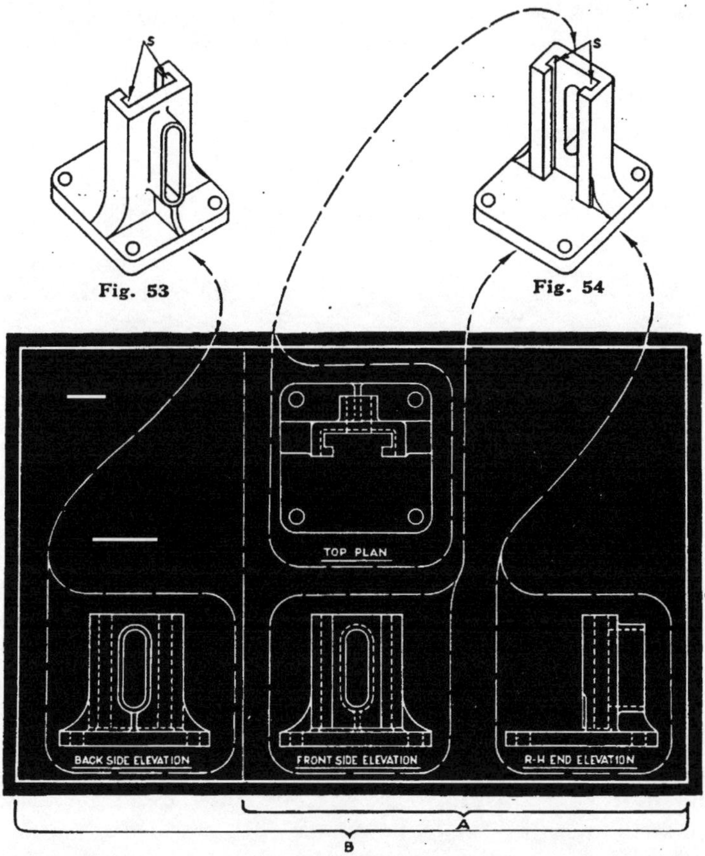

Fig. 53

Fig. 54

Fig. 55

to dotted or from dotted to full lines when viewed separately in each of the side elevations.

You will note in the front and back (side) elevations of section "B" of the blue print drawing of Fig. 55, that the lines forming the construction of the slot

"S" were not transposed from dotted to full lines, showing that those dotted (hidden) lines form the inside construction of the slot "S." The dotted lines of the slot "S" would mix up with those of the back construction in a confusing way ,if the front (side) elevation alone were shown as in section "A" of Fig. 55. The slot "S" as shown in the front (side) elevattion can be seen, of course, in the top plan.

THREE DETAIL BLUE PRINT DRAWINGS

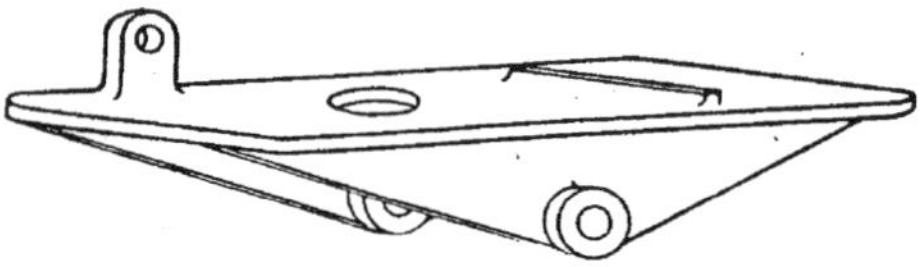

Fig. 56

The **three detail blue print drawing** as shown in Fig. 57 with two elevations and a top plan, is the kind most generally used. Such a blue print drawing gives three sources of information.

The word **"detail"** as used in connection with blue print drawings means simply the portion of the blue print drawing that an elevation or plan shows, like the top, the side or end elevation when considered separately from the whole blue print drawing. Each separate elevation or plan is known as a **detail.**

Almost any number of details may be shown in a blue print drawing, all depending upon the nature of the article drawn, which generally ranges from one, to as high as six details, and sometimes even more in an ordinary blue print drawing.

A **three detail blue print drawing** is generally what every blue print drawing shows, so as to give enough detail of what is seen in each side of the article drawn.

Figure 57 offers two elevations and a top plan for a blue print drawing, which is all that is needed to give the necessary information for the perspective drawing of Fig. 56.

A L-H. end elevation added to Fig. 57 would only show the Lug in a position opposite to that in which it is shown in the R-H. end elevation. Furthermore, such L-H. end elevation would necessarily have to be on the other side of the side elevation, where every L-H. end elevation belongs, and such would add no extra detail, other than what is shown in the R-H. end

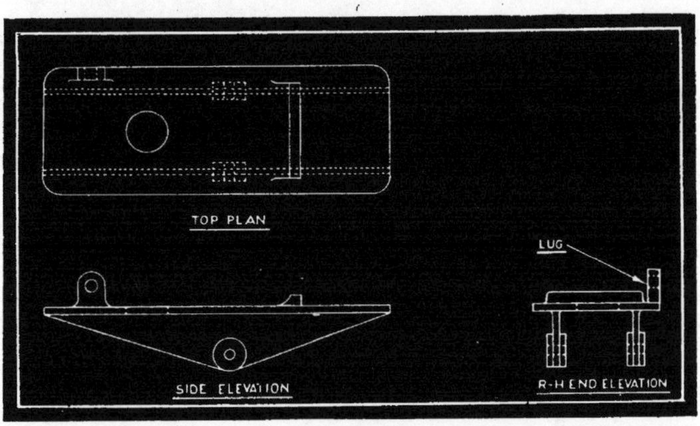

Fig. 57

elevation of the blue print drawing. Therefore, a draftsman would not use his valuable time in making an extra detail, which would be useless.

A bottom plan placed also in the blue print drawing of Fig. 57, with what is now shown, would not add any additional detail for a better understanding, other than what the top plan now shows. The only differerence in having a bottom plan shown, is that in **projecting** the bottom plan to the plane, there would be shown dotted (hidden) lines that are now seen in the top plan in full lines. Likewise, what is now seen in the top plan in dotted lines would be shown in a bottom plan in full lines, with the exception of the circle opening, which would be the same in the bottom plan, as now seen in the top plan.

Figure 59 is a blue print drawing of Fig. 58, which also has two elevations and a top plan. The details shown give sufficient information of the construction of Fig. 58, without any other additional elevations or plans.

It will be timely to dwell at this time on Fig. 59, and to review what was learned of the hinging of the Box as treated and explained in straight view projection. Apply the **"hinging method"** of obtaining the R-H end elevation to the plane from the side elevation of Fig. 59. What is your understanding of the instructions you have received so far of straight view

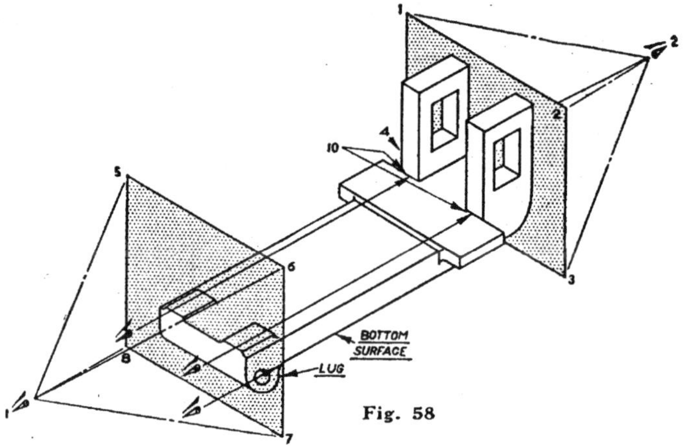

Fig. 58

projections? What end of the side elevation of Fig. 59 is it that **projects** in back of the R-H end elevation? The answer is, that the end as shown in the plane 5-6-7-8 of Fig. 58 is the end of the side elevation that lies in back of the R-H end elevation, and that the end of Fig. 58 that lies in the plane 1-2-3-4, is the end that shows in the R-H end elevation of Fig. 59.

From the last paragraph can be seen the real meaning of the planes 1-2-3-4 and 5-6-7-8 of Fig. 58. This should be a very good reminder of how the R-H end elevation was **projected** to plane from the side elevation of Fig. 59.

Should there be a L-H end elevation of Fig. 59 blue print drawing, what is seen through the plane 5-6-7-8 of Fig. 58 would be what is seen first in the L-H end elevation, and all that lies straight in back of the plane 5-6-7-8 would then be **projected** up to the plane of Fig. 59 to form a complete L-H end elevation.

If a L-H end elevation were shown in the blue print drawing of Fig. 59, everything that is seen of the

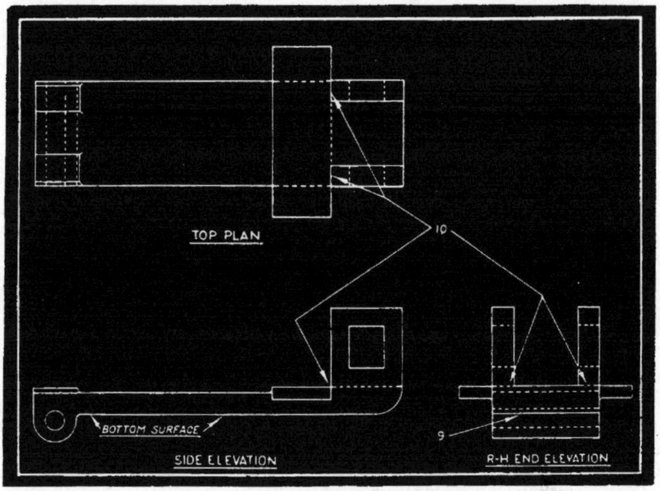

Fig. 59

R-H end elevation as shown would be the same for the L-H end elevation, with the exception of line number 9. Line number 9 would be drawn dotted in the L-H end elevation, instead of being drawn full as in the R-H end elevation. This is because the Lug which is in front of the bottom surface cannot be actually seen in a L-H end elevation, but only imagined as seen.

The lines of the blue print drawing of Fig. 59 as shown with the number 10, should help you to find where those lines as shown in one elevation or plan can be seen where they belong in each one of the other elevations or plans.

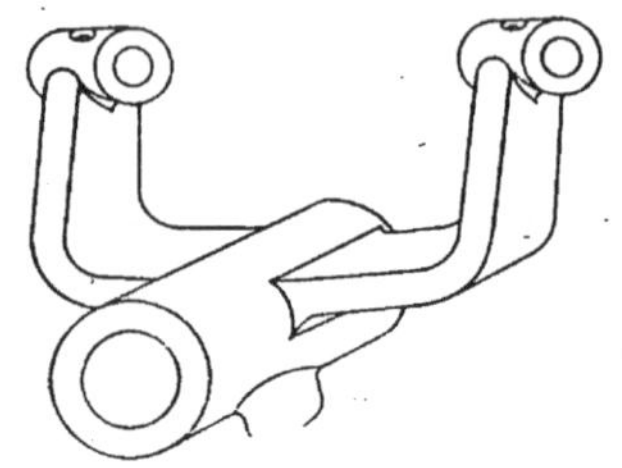

Fig. 60

Figure 61 is a blue print drawing which has two elevations and a top plan to be used in the making of the object shown in the perspective drawing of Fig. 60.

You will note that the arrangement of each elevation and plan of Fig. 61 is out of the regular order for details, which makes it different from those shown so far, with the exception of what has been explained of Figs. 16, 17, 18 and 19.

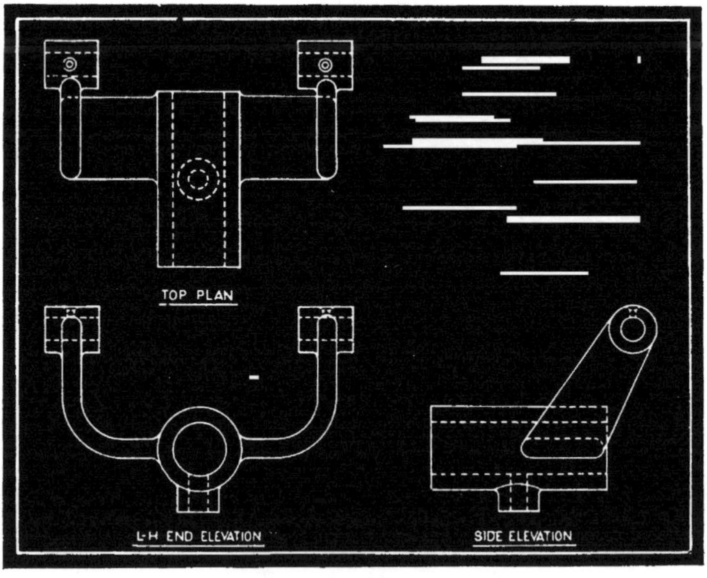

Fig. 61

The arrangement of the elevations and plan of the blue print drawing of Fig. 61 is correct in every respect, but if desired, the end elevation could be shown as a top plan, or a side elevation. Likewise, the top plan could be shown where the end elevation is now located. The end elevation would then be placed where the top plan is now placed. Such an arrangement is hardly ever made, because a top plan placed differently, would not be shown as the top of the article in its natural position. Should the elevations and the plan of Fig. 61 be so located, then the elevations and the plan would become different elevations and a different plan by the position they occupy in the whole blue print drawing.

TWO DETAIL BLUE PRINT DRAWINGS

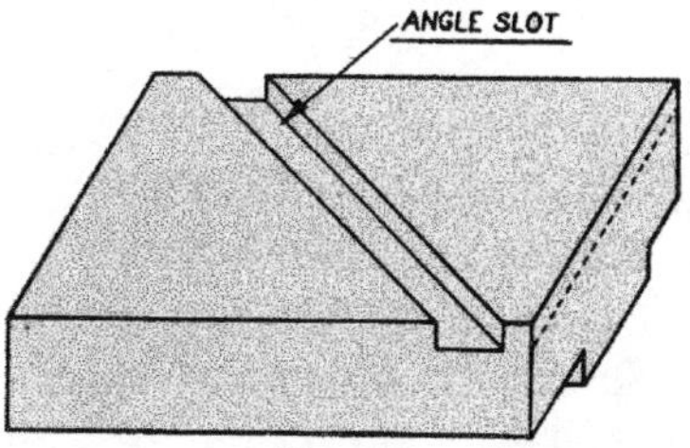

Fig. 62

Figure 63 is a blue print drawing containing a top plan and a R-H end elevation which is all the information that is needed to show the construction of the object in the perspective drawing of Fig. 62.

The top plan alone of the blue print drawing of Fig. 63 would not convey the necessary information concerning the object of Fig. 62 without some other elevation to work with it. A R-H end elevation therefore is provided, although a L-H end elevation or a side elevation could be placed with the top plan instead of the R-H end elevation as shown, and just as much information can be obtained from such an arrangement.

To show a bottom plan, instead of the top plan, as in Fig. 63, would require reversing what is now shown in the top plan, that is, the angled full lines

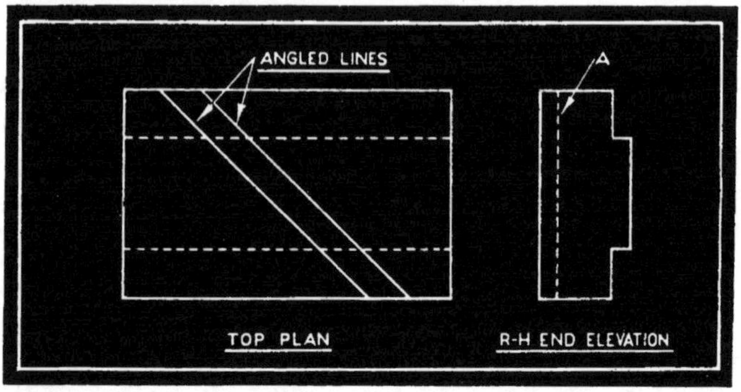

Fig. 63

of the top plan of Fig. 63 would have to be drawn dotted, and the dotted lines as shown in the top plan would have to be drawn with full lines in a bottom plan. Besides the turning around of the R-H end elevation relation to a converted bottom plan, if it was arranged that way.

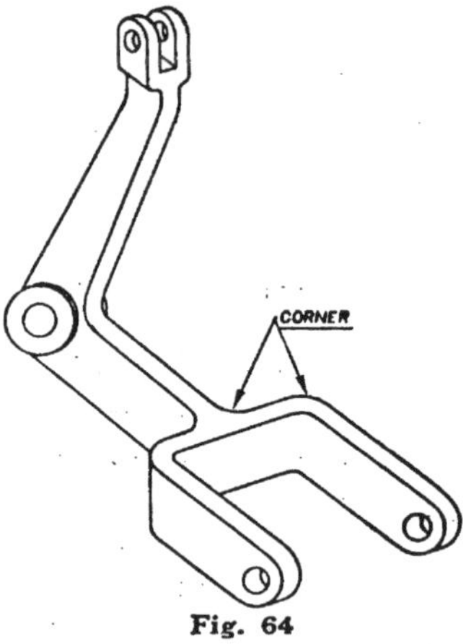

Fig. 64

A side elevation of the blue print drawing of Fig. 63 is not needed, because such information can be had by looking at the top or bottom side of the R-H end elevation, from which can be imagined a side elevation of Fig. 63.

The angled slot as shown in the top plan can be seen in the R-H end elevation for its depth, because in the R-H end elevation, the dotted line "A" shows how deep the slot is from the outside surface. The dotted line of the end elevation of Fig. 63 is suggested in Fig. 62 with a dotted line.

In Fig. 65 is represented with two details for a blue print drawing of an Off Set Yoke Bell Crank as shown in Fig. 64.

The blue print drawing of Fig. 65 is not complete in every detail concerning the Yoke Bell Crank. The shape of the corners is not shown in the blue print

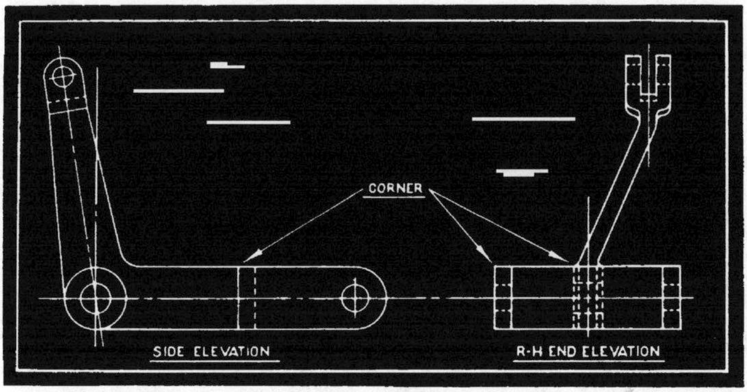

Fig. 65

drawing of Fig. 65. Of course Fig. 64 shows that they are round, but Fig. 64 is not drawn on a blue print drawing, and Fig. 65 does not show that the corners are round or square.

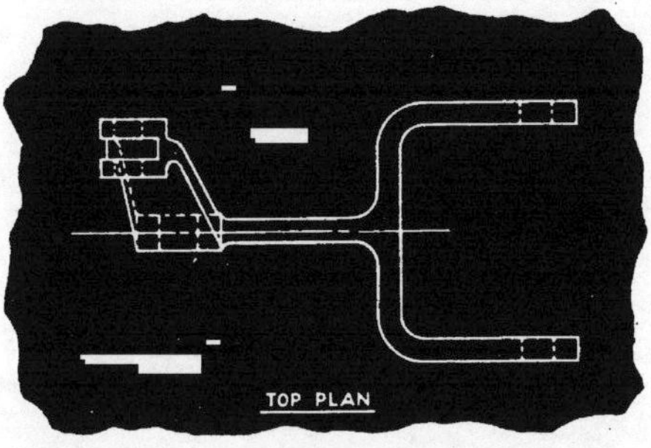

Fig. 66

With a **three detail blue print drawing,** there would have been no question of understanding for lack of detail.

A top plan like Fig. 66, given a place on Fig. 65 blue print drawing would convey all the necessary information for the corners in question, but this would be the only need of such a top plan in Fig. 65.

Should the side elevation of Fig. 65 be drawn with shaded lines, as Fig. 67 shows, only a **two detail blue print drawing** like Fig. 65 is all that would be needed. Figure 66 as a top plan in Fig. 65 would not then be needed, for the shaded lines in Fig. 67 would show the roundness of the corners. Since the amount of the roundness of the corners would have to be guessed at, the top plan of Fig. 66 in its place with Fig. 65 could be used to good advantage.

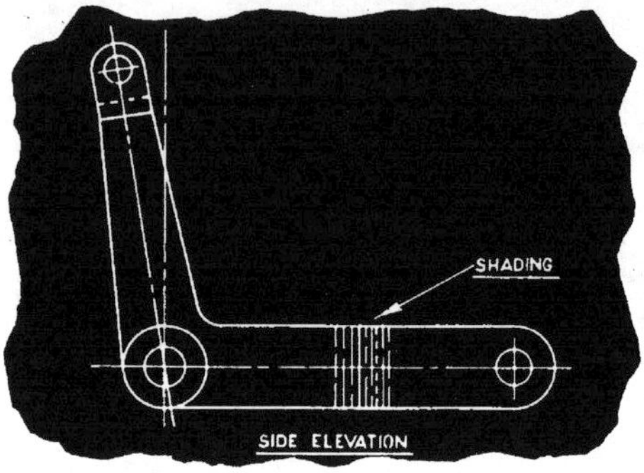

Fig. 67

Such shading lines as has been dwelt on in the above paragraphs will be more fully treated on in the subject of "CONVENTIONS."

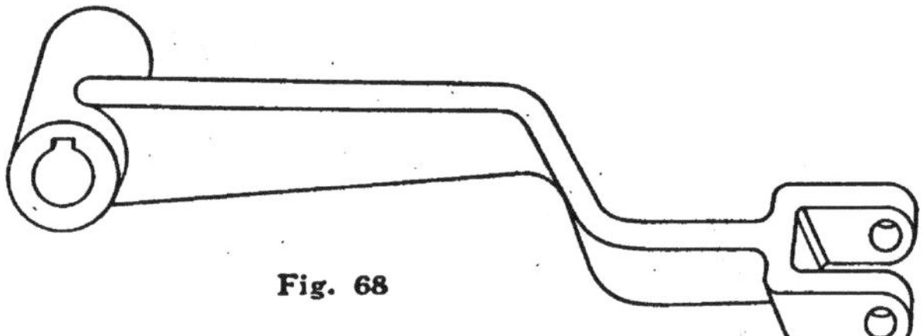
Fig. 68

Figure 69 represents with two details, a blue print drawing of the Off Set Lever Arm shown in Fig. 68.

With the top plan and side elevation that make up the blue print drawing of Fig. 69, there is not shown an end elevation, because an end elevation can be im-

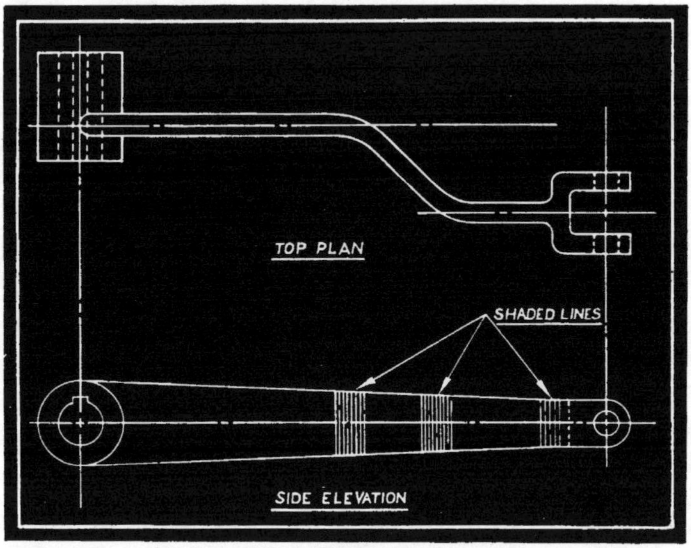

Fig. 69

agined from what is seen of the top plan and the side elevation. The addition of an end elevation to Fig. 69 would not help in any way.

ONE DETAIL BLUE PRINT DRAWINGS

One detail blue print drawings are not used very often, because they lack detail, and such **one detail blue print drawings** are limited to general application.

Fig. 70

Each of Figs. 70 to 78 could be drawn with end elevations, but such are not needed, for all necessary information is contained within these **one detail blue print drawings.**

As the **one detail blue print drawing,** as shown and treated on at this time, is so closely connected with the subject of "CONVENTIONS" that will be treated later, we will now only lightly touch upon the subject of **one detail blue print drawing.** It is necessary to take the subject up now, as **one detail blue print drawings** are often used to show a complete blue print drawing.

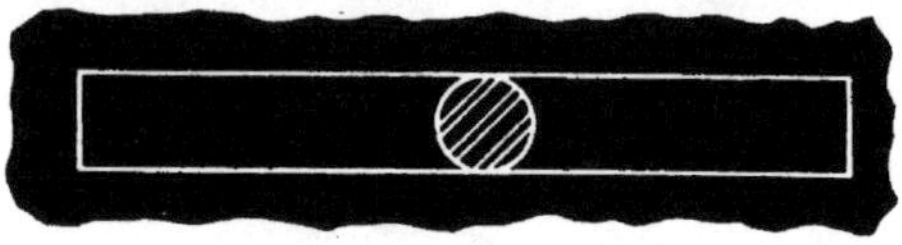

Fig. 71

Figure 71 is a blue print drawing of a round Bar, as shown in Fig. 70. If this Bar is not too long, a round circle is cross hatched inside of same to show that it is round. Should a **one detail blue print drawing** of Fig.

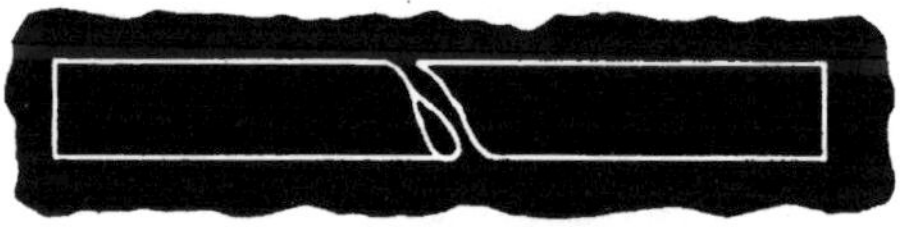

Fig. 72

71 be shown without the round cross hatched circle, it would not be understood without an end elevation. The mechanic would not know whether it was intended to be round or square.

Fig. 73

A cross hatched circle as shown in Fig. 71 is called an inserted cross hatching, which will be treated further in the subject of "CROSS SECTIONS."

Should Fig. 71 be extra long, too long to be drawn on a blue print drawing, then a break is shown in the Bar, like that of Fig. 72, which by its shape would show that the Bar is round, hence no round cross hatched circle like that in Fig. 71 is needed to show that the Bar is round.

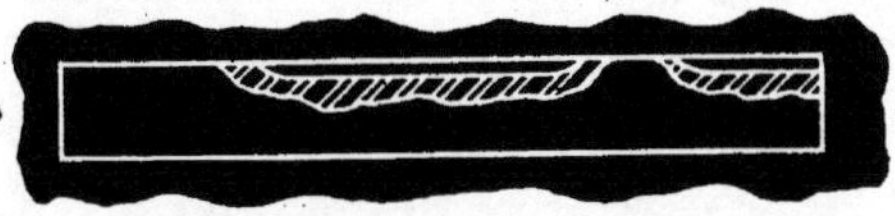

Fig. 74

Figure 74 is a torn out part of a blue print drawing of a Shaft, with keyways placed as shown in Fig. 73. The top part of the Shaft of Fig. 74 is shown broken out, so that cross hatching may show a separation of the metal that the Shaft is made of, from that of the position of the keyways.

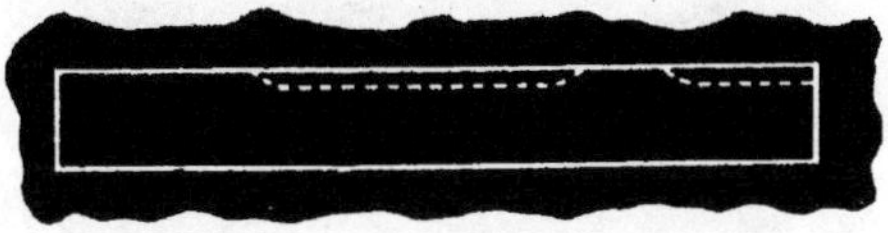

Fig. 75

Figure 74 shows the regular way of representing keyways in Shafts. Keyways are used only on Shafts, and since all Shafts are round, no other lines are needed. If a blue print drawing is made to represent an object other than a round Bar, with a keyway in it, then that object is a different shape and considered special, and for special instruction, which can be given in detail by means of another elevation or plan.

Fig. 75 is a blue print drawing of the same construction as Fig. 74, but instead of showing the keyway within a broken out section, the keyways are drawn dotted, because the dotted lines shows that the construction is not open to full view.

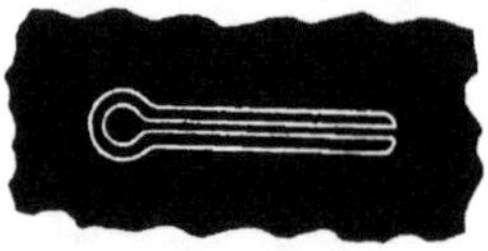

Fig. 76

Figure 76 is a Cotter Pin known to all machine tradesmen. A drawing as in Fig. 76 is all that is necessary, because of the article's commonness. Its construction must not be of any other nature than along the manufacturers' standard, otherwise, another elevation or plan would be necessary to show its special construction.

Figure 77 shows a **one detail blue print drawing** of a Ball. The one detail is all that is necessary to represent the Ball. The circle drawn without the shading, as Fig. 78 shows, would convey no meaning, unless a note were placed under same to give further information as to the meaning of the circle.

An end elevation or a top plan of the Ball placed beside the side elevation of Fig. 78, would make a

Fig. 77

Fig. 78

group of circles, which circles would naturally be of the same size and would cause confusion. Therefore, if a Ball is to be drawn for a blue print drawing, shading as in Fig. 77 must be made to show that it is a Ball.

A mere circle to represent a blue print drawing of a Ball, as in Fig. 78, would not be understood, for such a circle could be thought of as a circle of a plate.

HALF DETAIL BLUE PRINT DRAWINGS

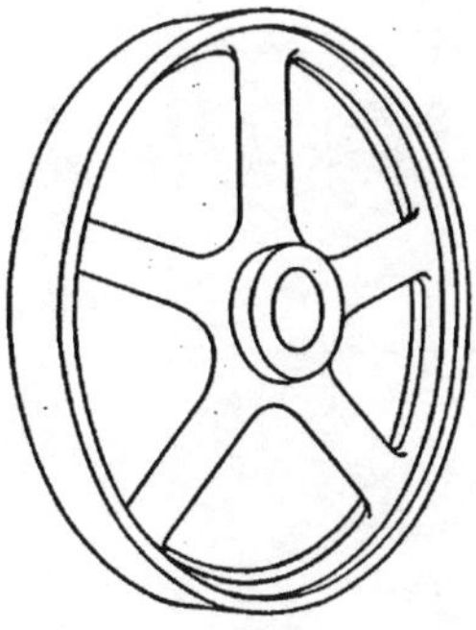

Fig. 79

The purpose of making a **half detail blue print drawing** would not apply for all classes of objects, but only to such objects as a Wheel, Pulley, etc., which are made up of two parts exactly alike.

The purpose in drawing a **half detail blue print drawing** such as that shown in Fig. 80, is to save work by making only half of the drawing rather than the whole.

Another reason why a **half detail blue print drawing** is used, is because a **half detail** may be drawn larger, so as to permit larger space for details which could not be shown in the cramped space of a smaller drawing.

It must be understood that a **half detail blue print drawing** can only be made of an object whose two halves are exactly alike.

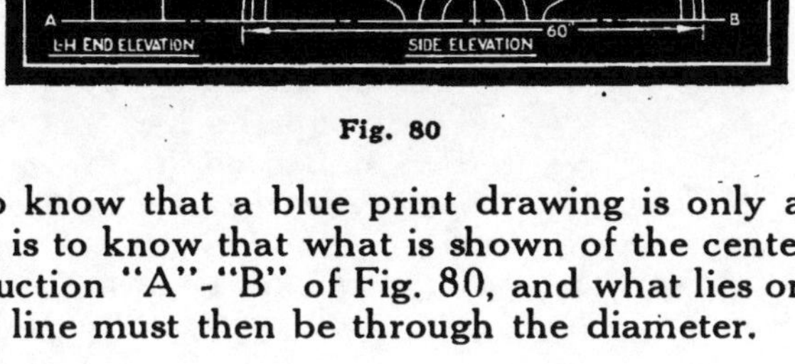

Fig. 80

To know that a blue print drawing is only a **half detail,** is to know that what is shown of the center line construction "A"-"B" of Fig. 80, and what lies on that center line must then be through the diameter.

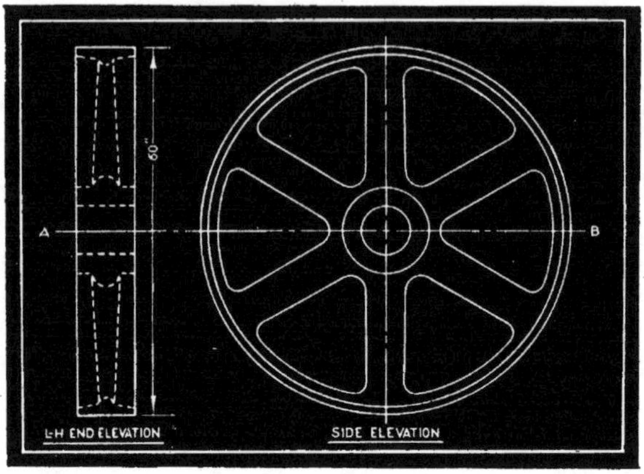

Fig. 81

Figure 81 is a complete blue print drawing of Fig. 79 that was shown drawn with a **half detail blue print drawing** in Fig. 80.

In Figure 83 is shown a **half detail blue print drawing** similar to that explained in Figs. 80 and 81, the only difference being that Figs. 80 and 81 have six spokes, while Figs. 83 and 84 have only five spokes.

Figure 84 is a complete blue print drawing of a five spoke Wheel which was shown drawn with a half **detail blue print drawing** in Fig. 83.

Fig. 82

It was said in the explanations for Fig. 80, that in order for a **half detail blue print drawing** to be drawn, the half that is drawn, must be like the half that is not drawn, but in Fig. 83, the half that is drawn, is not like the half that is not drawn, because it shows three spokes in the top half of Fig. 83, while the bottom half would only have two spokes. This condition is taken care of with a note above the detail of Fig. 83, stating: "NOTE—5 SPOKES

EQUALLY SPACED," which note meets the objection of not showing each half, when the halves are unlike.

Figure 84 is a complete blue print drawing of a five spoke Wheel that was shown drawn with a **half detail blue print drawing** in Fig. 83.

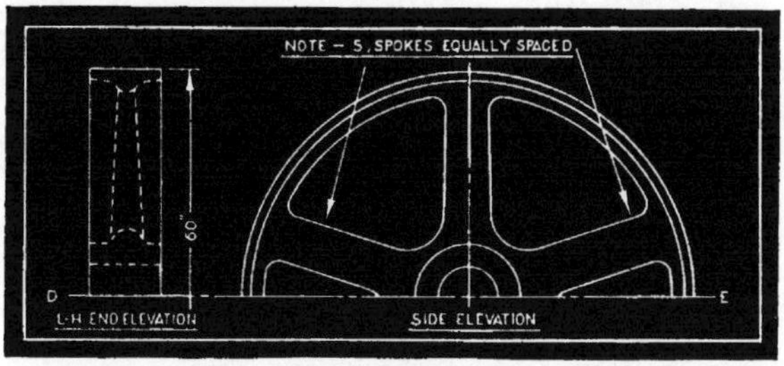

Fig. 83

The dimensioning of the diameter of the **half detail blue print drawing** can be understood in Fig. 80, for its diameter is shown, but most generally in such **half de-**

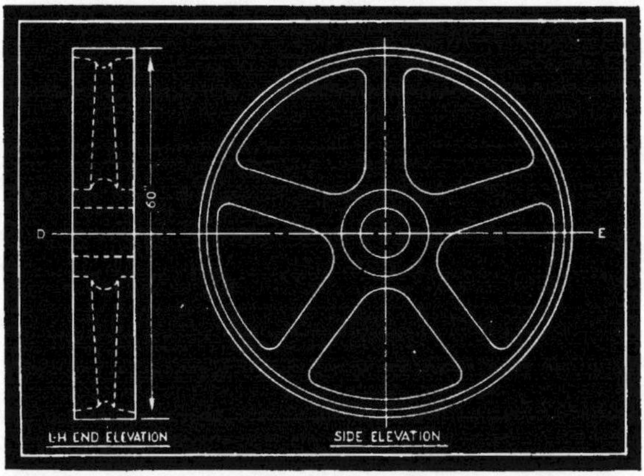

Fig. 84

tails, the diameter is shown like in the end elevation of Fig. 83, with the dimension line extending past the center line "D"-"E" which is to be understood as the whole diameter.

PART V

ANGLE PROJECTION ELEVATION

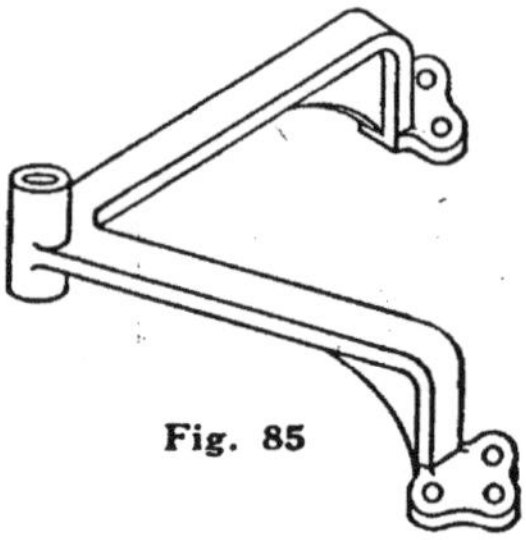

Fig. 85

Almost every blue print drawing is of the straight projection kind, but often a blue print drawing, like Fig. 87, has an **angled projection.**

Figure 87 shows a complete **angle projection** blue print drawing, with all the details of elevations and plans that are necessary in representing the Arm Bracket shown in Fig. 85.

The method of **projecting** the side elevation to the plane from the top plan, in Fig. 87, is somewhat different from the general run of blue print drawings.

First:—The side elevation as shown **projected** to the plane of Fig. 87 is where the end elevation is generally located in almost all blue print drawings. This does not matter, as was explained before, because any **projection** can be made from any elevation or plan.

Second:—The side elevation as **projected** to the plane from the top plan of Fig. 87 is shown drawn in an angled or slanted position is explained on the principle of the triangle shown in Fig. 86.

In Fig. 86 is shown a right angle triangle. The undeniable law of the right angle triangle is: That the hypotenuse is opposite the right angle which is the longest side (or leg). Should the triangle not be a right angle triangle, then the side opposite the greater angle of the triangle is the longest side of the triangle.

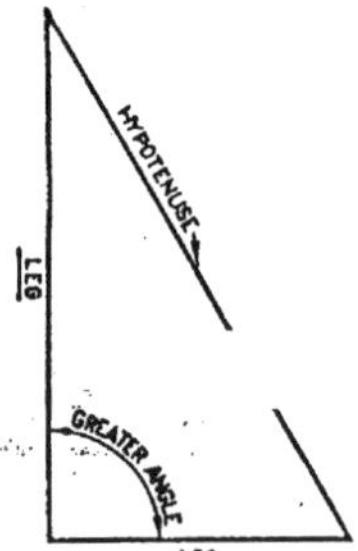

Fig. 86

In the top plan of Fig. 87, you will find two triangles, one numbered "1-2-3" and the other "1-3-4," having the line "1-3" as the one common hypotenuse for both of the triangles. These two triangles are butted or placed together, making with the two triangles, the rectangle "1-2-3-4."

The side elevation of Fig. 87 is **projected** to the plane from the top plan in an angled or slanted position, which in this case is the proper position for the side elevation to be **projected** from the top plan, as shown in the blue print drawing of Fig. 87.

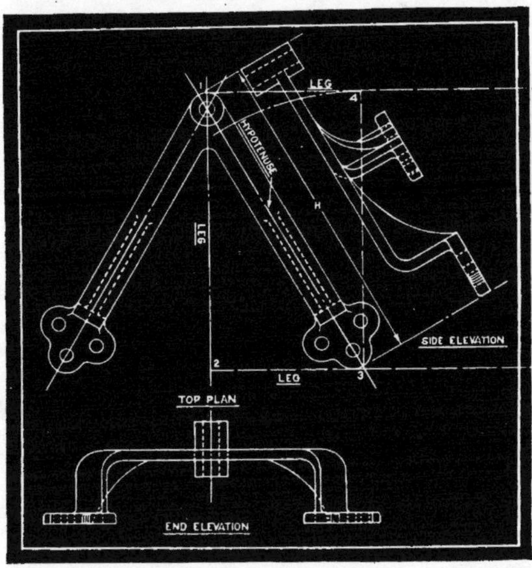

Fig. 87

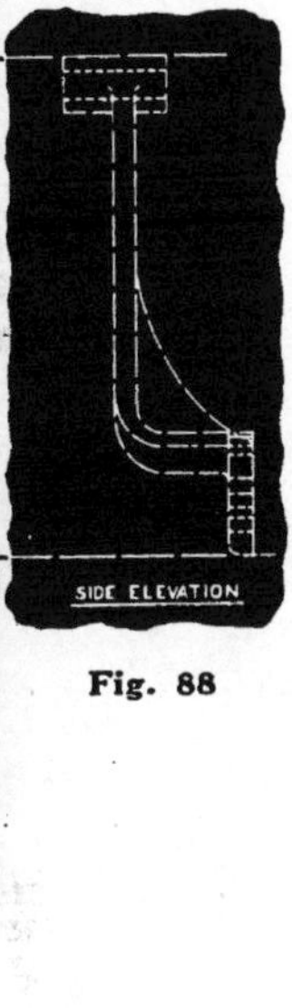

Fig. 88

The side elevation of Fig. 88 is shown drawn with dashed lines, as the way almost all blue print drawings show their side elevation **projected** position. But the side elevation of Fig. 87 is not drawn on the same principle as the side elevation of Fig. 88. The side elevation in Fig. 88 is not properly **projected,** because it is **projected** from the leg "3-4" of the triangle, which is shorter than the hypotenuse "1-3," which is not the true length of the Arm of the Arm Bracket, where as the side elevation of Fig. 87 is **projected** from the top plan from the hypotenuse "1-3," which is the true length of the Arm of the Arm Bracket.

The explanation in the last paragraph makes clear that the side elevation of Fig. 87 **is correct,** and that the side elevation as shown in Fig. 88 **is not correct** for a blue print drawing where there are elevations or plans **projected** from an angled object, such as the top plan of Fig. 87 shows.

PART VI

CROSS HATCHINGS

Cross hatching is made of several slanted lines grouped into a **design** or **symbol** that has a general meaning to those who understand the reading of blue print drawings, as the **symbols** of Fig. 89 show. When any one of these **symbols** is drawn into a cross section, like that in Fig. 91, it represents the material to be used in making the articles drawn.

Each cross section may show with one or more of the several styles of the slanted **symbols,** as in Fig. 89, what is desired of what the blue print drawing show.

Some draftsmen vary in degree on one or more of the several **symbols of materials** as shown in Fig. 89. Some may use a **symbol** to convey a different meaning than that is shown in Fig. 89, but these differences are generally done by those draftsmen who are more or less lacking in experience and practice.

Figure 89 does not, however, show every **symbol** that may be used at times, for a **symbol** may be created by a draftsman to show some special materials.

Should you not readily understand what a **symbol of material** means in a blue print drawing, you can always find out, for the name of the material is also generally shown below the elevation or plan of the detail, but such material is never stated in an assembly blue print drawing.

When **symbols** of **cross hatching** of Fig. 89 are used in an assembly blue print drawing, it is to principally show the separation of one or more parts that compose the whole assembly, besides what the cross section is to show.

There is no rule as to how far apart these slanted **cross hatch lines** should be, as in Fig. 89, but it is

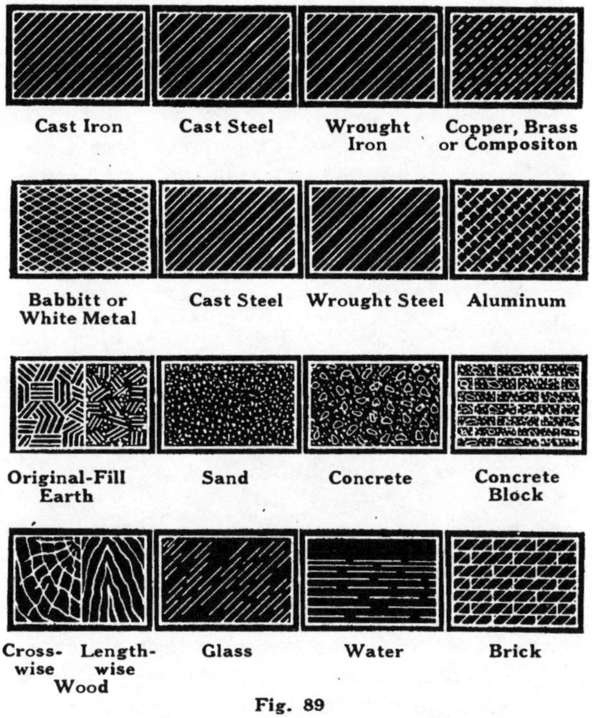

Fig. 89

Symbols of cross hatching.

expected that the spacing of **cross hatchings** will be equally placed apart, according to the size of the space that the **cross hatching** occupies.

The reader should memorize and know at sight,

the different **symbols of material,** so that when he sees a **cross hatching** in a cross section, he understands what the material is that is represented in the blue print drawing.

CROSS SECTIONS

This chapter on **cross sections** is of great importance, for it enables the reader of a blue print drawing to understand many special facts not obtained otherwise.

Should a blue print drawing that you are to read be made up by a capable draftsman, that draftsman will make use of **cross sections** quite freely, so as to make as far as possible his drawings self-explanatory. One who is versed in the fundamental principles of blue print drawings will know what is meant with but little study and effort.

The main rule to follow in making a **cross section** of a blue print drawing, is to show what is to be imagined of the detail of the object, if it were cut into two parts, so as to show the shape of the section that is in view when the parts are separated.

A **cross section** of any blue print drawing is the only place where a cross hatching (as explained in the last chapter) can be used, that is why a **cross section** is always known at a glance. The **design** or **symbol of material** that is to be used is at once noticeable.

Cross sections are not shown in every kind of a blue print drawing, because many blue print drawings are simple in design and need no extra helps to the mechanic. **Cross sections** are most generally used in elevations and plans to show the shape of a corner or thickness. They are used especially when an elevation of a blue print drawing is complicated in shape.

All **cross sections** of any blue print drawing are always taken from a line, which is most generally from the center line of the elevation or plan that is next to the elevation or plan in which the **cross section** is shown. In other cases, a **cross section** may be taken from a **created line** to show where a cross section is desired.

There are various classes of **cross sections,** which are:

Center line cross sections
Off center line cross sections
Quarter cross sections
Extended out cross sections
Broken out cross sections
Inserted cross sections
Zig zag cross sections
Rib cross sections
Created line cross sections
Assembly cross sections.

Each of the above **cross sections** will be explained in the same order as here listed.

CENTER LINE CROSS SECTION

In Fig. 91 is a blue print drawing of a Wheel, the end elevation of which shows a **center line cross section.** This Wheel is shown by the use of the side and end elevations, which are all the elevations that are needed for a blue print drawing of the Wheel, because the addition of any other elevation or plan would add nothing to the meaning of the detail of the object to make it clearer, other than what the blue print drawing of Fig. 91 shows.

The end elevation of the Wheel in Fig. 91 shows a cross section which was taken from the center line "A"—"B" of the side elevation. This end elevation cross section of Fig. 91, like all cross sections of any blue print drawing, is always looked for on the center line of the elevation that is next to the end elevation that is cross sectioned.

The center line "A"—"B" which passes thru the side elevation of Fig. 91, and extends as the same center line "a"—"b" thru Fig. 90, is that which you see cross hatched in the end elevation of Fig. 91.

The half of Fig. 90 that is cross hatched, is to be imagined as that seen in the center line of the side elevation of Fig. 91, and given in detail in the end elevation.

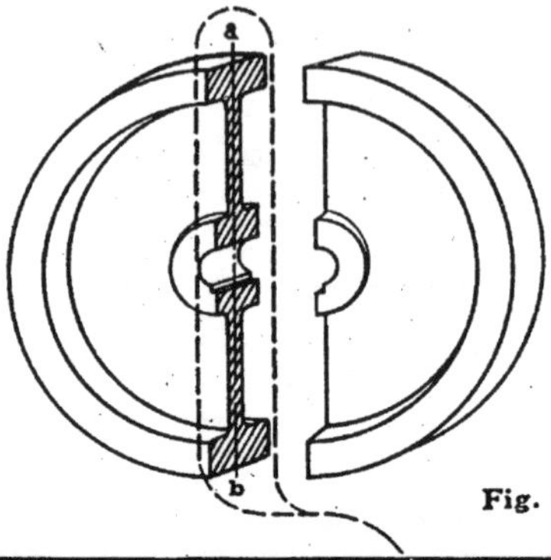

Fig. 90

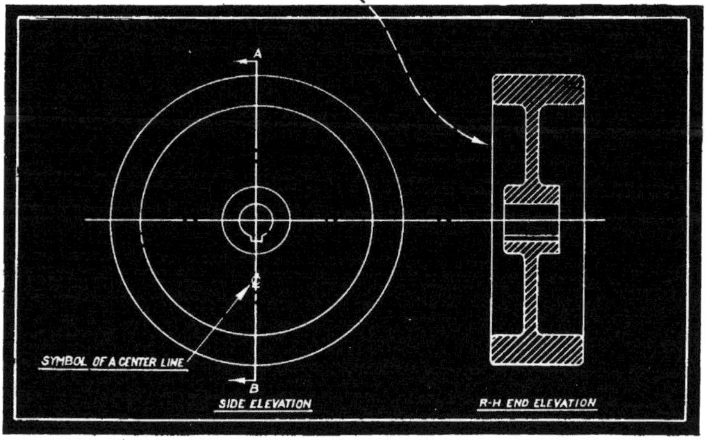

Fig. 91

To make the last paragraph apply as an explanation of all **center line cross sections** of every detail of any blue print drawing, train your mind to imagine that what is seen thru the center line of the side elevation, if same were cut in half, would be that as shown in Fig. 90 **cross sectioned end.**

What has been explained of Figs. 90 and 91, applies exactly to Figs. 92 and 93, with the exception that the end elevation of Fig. 93 is not a cross section.

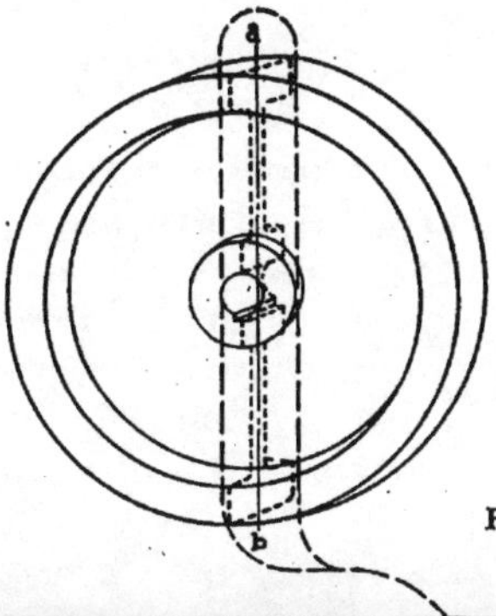

Fig. 92

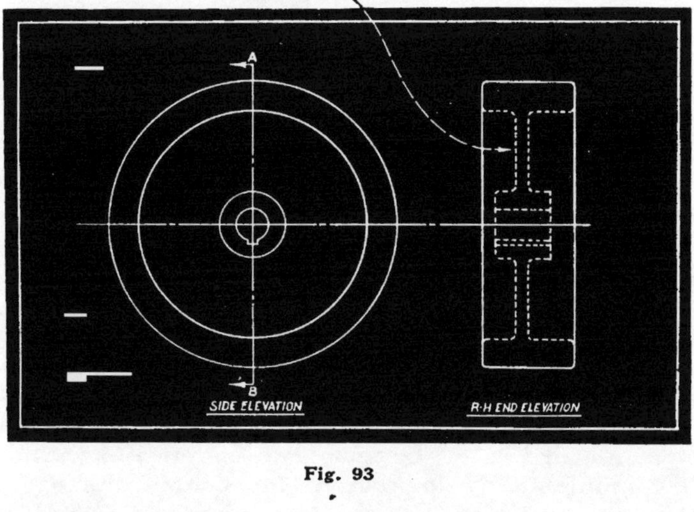

Fig. 93

Figures 91 and 93 are drawn for the purpose of comparing an elevation that is cross sectioned with one that is not.

In Fig. 93, the vertical center line "A"—"B" that passes thru the center of the side elevation, also passes thru the dotted section of the web of the Wheel in

Fig. 92. This dotted section of the Wheel is the same as that **projected** by the use of dotted lines from the center line of the side elevation to the end elevation of Fig. 93.

The dotted section as shown in the end elevation of Fig. 93 is that half of the side elevation of the Wheel that is shown by the arrows directed at each end of the center line of the Wheel of Fig. 93.

It is seen from the above paragraph, that it is not necessary that an end elevation of any blue print drawing should be cross sectioned, only that cross sections assists and helps the reader of a blue print drawing to a better understanding, especially should the drawing be complicated in nature.

In comparing Figs. 91 and 93, each figure is drawn correctly, and either one or the other of the Figs. 91 and 93 could be drawn, as each may be used alone with but little difference as to the possibilities of it being understood. It is most likely that Fig. 91 would be the choice of a draftsman in drawing a Wheel.

OFF CENTER LINE CROSS SECTION

In explaining Figs. 90 to 93, it was stated that when any cross section is made on any elevation or plan of a blue print drawing, that it is generally found, and may be looked for on a center line of one of the elevations or plans that is connected with the cross section of the object drawn.

Should you after seeing the cross section in the end elevation of Fig. 95 be unable to understand within a few moments by looking at the center line of the elevation that is next to the cross sectioned end elevation, of where the cross section comes from, you are then to look for remarks below the elevation or plan that is cross sectioned.

The remark "SECTION ON D—D" as stated below the cross sectioned end elevation of Fig. 95, should direct you to the line "D—D," and not to the center line of the side elevation where the cross section comes from.

By extending the line "D—D" in the side elevation of Fig. 95 straight up thru Fig. 94, would change the

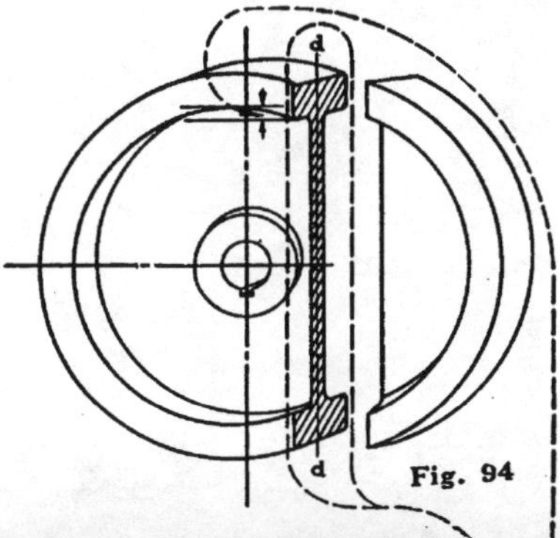

Fig. 94

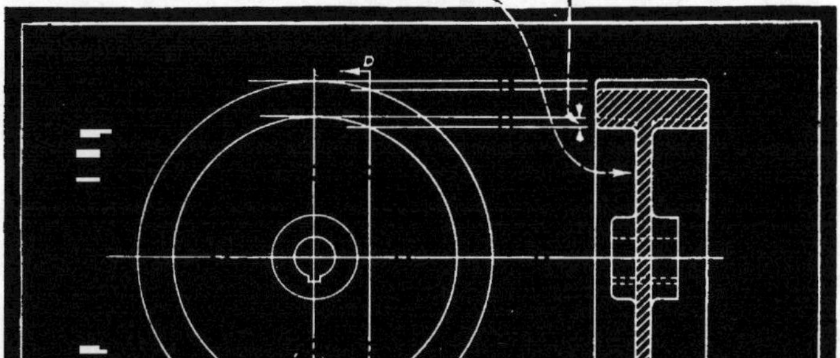

Fig. 95

off center line "D—D" of Fig. 95 to "d—d" of Fig. 94 from which the end elevation of Fig. 95 is to be realized.

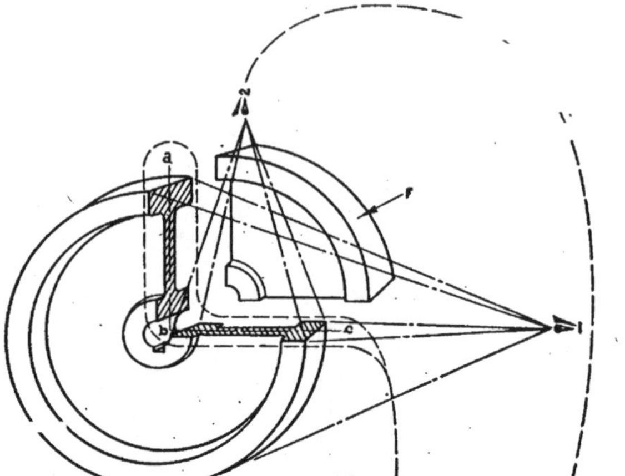

Fig. 96

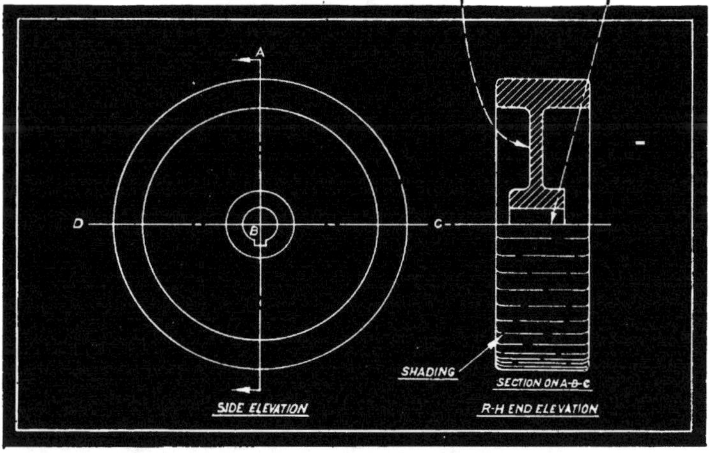

Fig. 97

QUARTER CROSS SECTION

This problem of **quarter cross section** is somewhat the same in principle as the cross sections already explained.

The **quarter cross section,** like the end elevation of Fig. 97, is drawn usually this way by draftsmen so as to gain time, and at the same time convey just as full meaning of the center line as the other cross sections already explained.

When a cross section like that shown over the horizontal center line "D-B-C" of the side elevation of Fig. 97 is not a complete cross section for the whole of the end elevation, then remarks as are shown below the end elevation of Fig. 97 are to be looked for, telling where in the side elevation that the cross section does come from.

Below the end elevation of Fig. 97 are the words "SECTION ON A-B-C." The end elevation is to be **projected** from the line "A-B-C" of the side elevation, which would show exactly what the eye could see when looking straight from the center of the vision lines of the eye position number 1 of Fig. 96, when the "F" section of the Wheel is imagined as taken away. What is seen from the eye position number 2 is not to be seen in the end elevation, but imagined as seen.

The cross section of the end elevation of Fig. 97 is generally considered by one who reads blue print drawings that the shape cross hatched for the top half, is to be the same shape as the bottom half, which is not shown cross hatched.

The shading as shown in the end elevation of Fig. 97 is to represent the roundness of the face of the bottom half of the Wheel, which is not cross sectioned.

EXTENDED OUT CROSS SECTION

In Fig. 98 is shown a complete blue print drawing of a Lever Handle, with the perspective drawing of Fig. 100 to help in the following explanations.

Extended out cross sections are cross sections shown **extended out** from an elevation or plan, with remarks stated below each, such as:—"SECTION ON A-B," and "SECTION ON C-D" as seen on the blue print drawing of Fig. 98, is what is **projected** from the lines "A-B" and "C-D" of the end elevation.

The **extended out cross sections** as seen in the blue print drawing of Fig. 98, is seen for its position in the Lever as shown in Fig. 100.

The **extended our cross sections,** as the "SECTION ON A-B" of Fig. 98, answers practically all that a top plan could offer for information of the Lever, there-

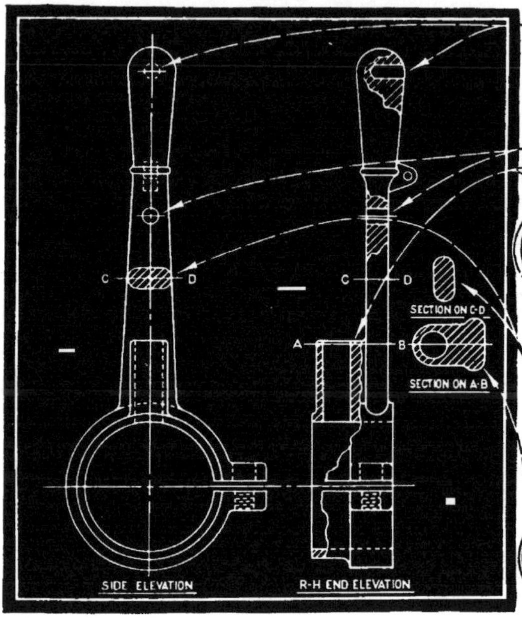

Fig. 98

Fig. 99

Fig. 100

fore a top plan is not needed for Fig. 98, but were the "SECTION ON A-B" not shown, then a top plan would be needed to show the roundness at the part "A" of the end elevation.

INSERTED CROSS SECTION

An **inserted cross section** is a cross section for a particular part or section placed inside of an elevation or plan.

The line "C-D" of the end elevation of Fig. 98 as seen extended out and cross sectioned is all that is needed to show a cross section of that particular part of the Lever, but should there not be an extended out cross section as shown, and should there be any cross section desired at that point of the lever, an **inserted cross section** could be shown inside of the side elevation instead.

When an **inserted cross section,** such as is shown upon line "C-D" in the side elevation of Fig. 98 is used, the extended out cross section "C-D" as extended out from the end elevation is not needed, for they both show the same detail.

BROKEN OUT CROSS SECTIONS

In the end elevation of the blue print drawing of Fig. 98 are shown three **broken out cross sections** which are to represent the parts of the Lever shown by the encased arrow lines of Fig. 99.

What is seen cross hatched in the **broken out cross sections** of the end elevation of Fig. 98, is to be understood as on the center lines of the side elevation, for, as was explained before, that all cross hatchings are most always shown on center lines, unless there are remarks given below the elevation of where the cross section did come from.

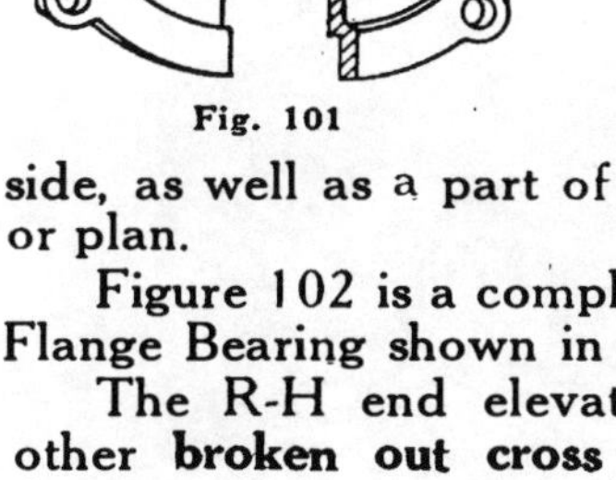

Fig. 101

The reason for showing **broken out cross sections** in Fig. 98, is to give a combination of the inside, as well as a part of the outside of any elevation or plan.

Figure 102 is a complete blue print drawing of the Flange Bearing shown in Fig. 101.

The R-H end elevation of Fig. 102 shows another **broken out cross section** for the purpose of showing in open view, the outside of the hole "B," and the inside of the part "A" of the Flange Bearing, without using dotted lines as in the end elevation of Fig. 103.

The **broken out cross section** of the R-H end elevation of Fig. 102 is not really necessary, for the R-H. end elevation of Fig. 103 could be shown instead.

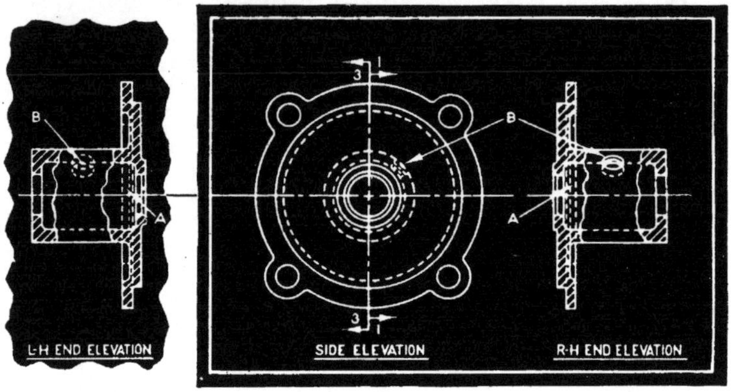

Fig. 102

The L-H end elevation as shown on the side of the blue print drawing of Fig. 102 is not needed, because the R-H. end elevation shows the **broken out cross section** of the outside of the hole "B," while in the

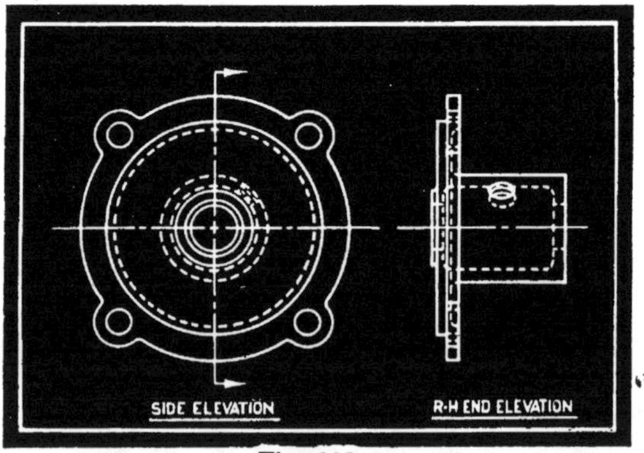

Fig. 103

L-H end elevation of Fig. 102, what is shown in dotted lines of the hole "B" has a hazy meaning.

Were all of the R-H end elevation of Fig. 102 to be cross hatched, it would be impossible to show the hole "B," from the direction of the center line arrows No. 3 of the side elevation direct, would not contain the hole "B," although the end elevation could be cross hatched to show the hole "B" if imagined projected from the arrows No. 1 of the side elevations.

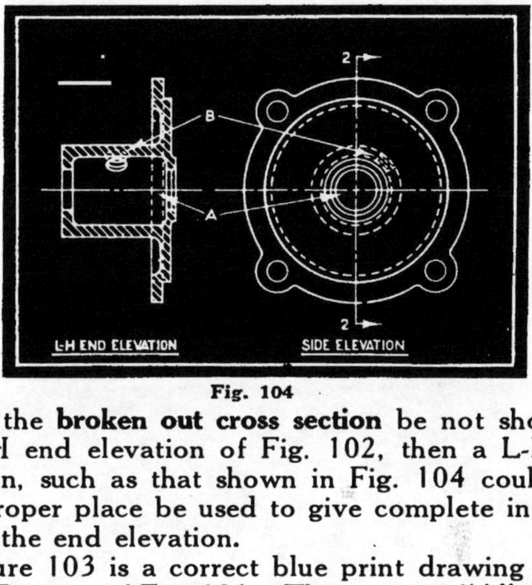

Fig. 104

Should the **broken out cross section** be not shown in the R-H end elevation of Fig. 102, then a L-H end elevation, such as that shown in Fig. 104 could also in its proper place be used to give complete information of the end elevation.

Figure 103 is a correct blue print drawing of the Flange Bearing of Fig. 101. The opening "A" as seen in Fig. 101 and shown with dotted lines in the end elevation of Fig. 103 is not as clearly understood as the same R-H end elevation of Fig. 102 would be with a **broken out cross section.**

The position of the hole "B" in the L-H end elevation of Fig. 104 is seen in its position, because the hole "B" as shown in the L-H end elevation is in the half of the side elevation of Fig. 104 marked by the arrows of the center line in the direction of No. 2.

ZIG ZAG CROSS SECTIONS

One of the chief advantages of the **zig zag cross section** in a blue print drawing is shown by the special Block of Fig. 106.

In Fig. 106 is shown a complete detail blue print drawing of the Block represented in Fig. 105.

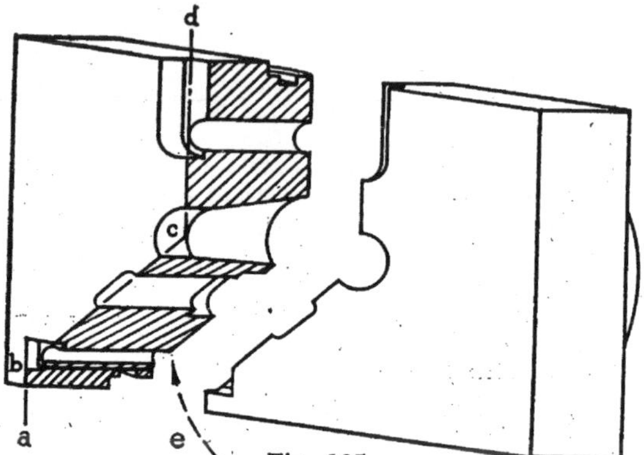

Fig. 105

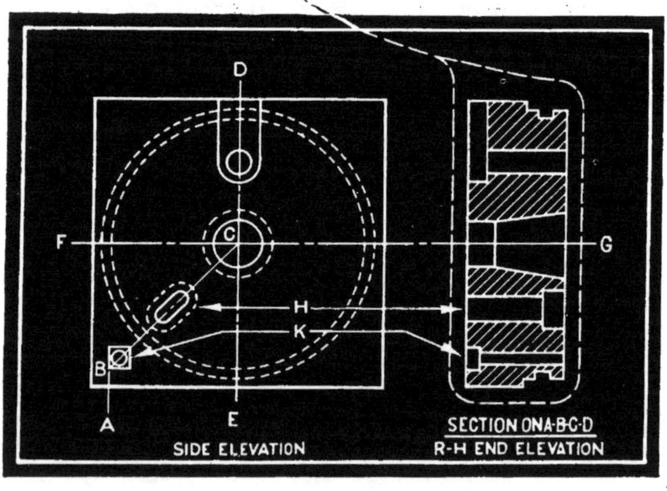

Fig. 106

Below the end elevation cross section of Fig. 106 are the words: "SECTION ON A-B-C-D," which means that the cross section was taken from the line "A-B-C-D" of the side elevation.

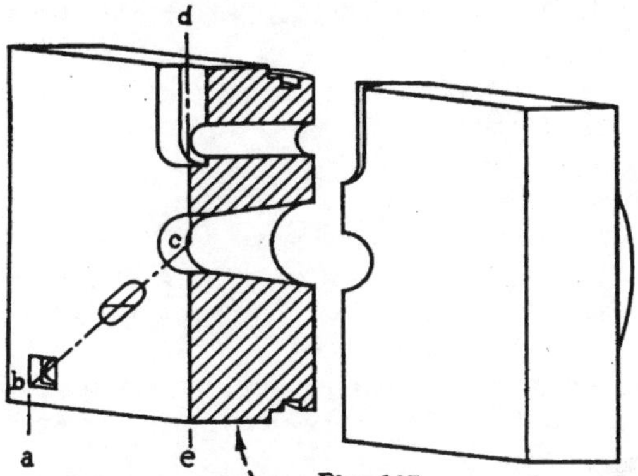

Fig. 107

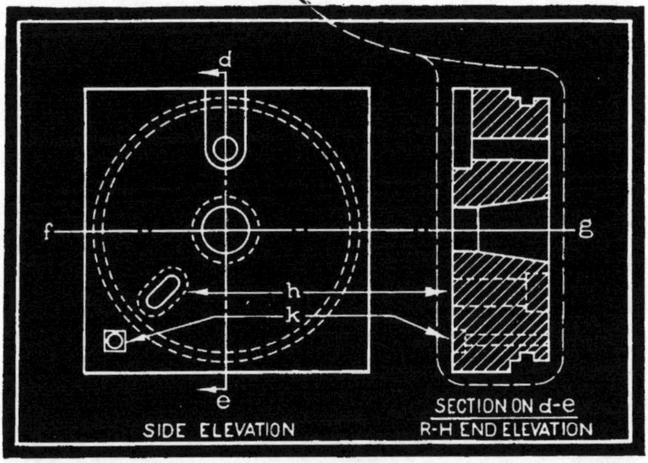

Fig. 108

What is shown cross sectioned in the end elevation of Fig. 106 is to be understood in the side elevation as that which is shown cross hatched in Fig. 105.

The advantage of the **zig zag cross section end elevation** as taken from the line "A-B-C-D" of the side elevation of Fig. 106, is that it shows all the shape of the holes and openings of the side elevation in full open view in the end elevation, but should the end elevation of Fig. 106 be taken from the center line "D-E" instead of from the zig zag lines "A-B-C-D," then the end elevation of Fig. 106 would be like that shown of the end elevation of Fig. 108.

In comparing Fig. 106 with that of Fig. 108, it will be found that both are drawn correct, but that the end elevation of the zig zag of Fig. 106 gives clearer detail of the "H" and "K" openings of the Block, than the end elevation of Fig. 108 does, because the end elevation of Fig. 106 is **projected** from the side elevation as that which is seen openly from "A" to "B," from "B" to "C" and from "C" to "D." In Fig. 108, the cross sectioned end elevation is taken from the center line "d"-"e" of the side elevation which is unlike the cross sectioned end elevation of Fig. 106, because in Fig. 108, the openings "h" and "k" are seen in the cross sectioned end elevation with dotted (hidden) lines as being back of the center line "d"-"e." The explanation for this is clearly shown for the relation of Fig. 105 to Fig. 106 and likewise in that of Fig. 107 is to Fig. 108.

In Fig. 110 is shown another example of the **zig zag cross section** in the blue print drawing of a Die Block. This Die Block is shown with three openings in its side elevation that could not appear on one common center line, because a center line cross section could be used in showing the openings back of the center line only in dotted lines, which would not then be a **zig zag cross section.** To show these openings so as to get a clear open section for the end elevation of Fig. 110, it is necessary to take a cross section along the **zig zag line** "A-B-C-D-E-F" of the side elevation.

The end elevation of Fig. 110 is a **zig zag cross section,** which below this end elevation are the words:

"SECTION ON A-B-C-D-E-F" as projected from the side elevation. The end elevation should be imagined as all of the cross sectioned end of the Die Block "M" taken apart on the **zig zag line** "a-b-c-d-e-f" in Fig. 109.

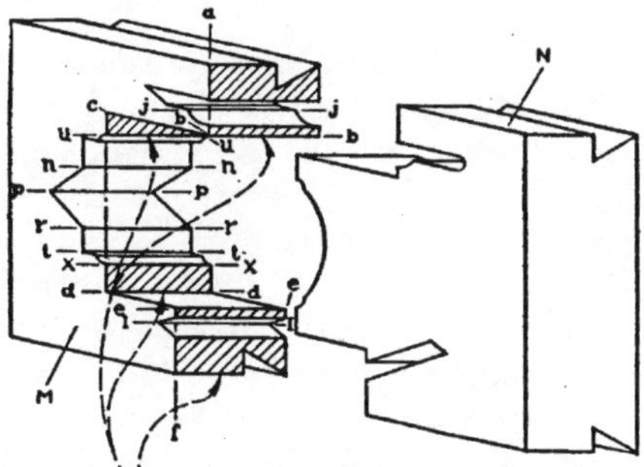

Fig. 109

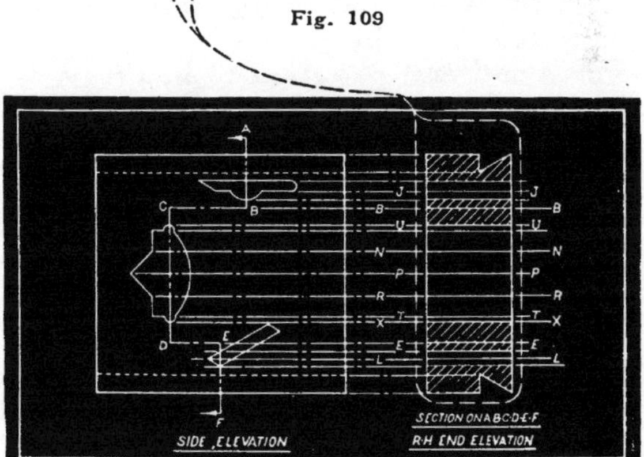

Fig. 110

All of the cross hatched ends of the Die Block "M" of Fig. 109 are the very same as those cross hatched in the plane of the end elevation of Fig. 110. These are to be thought of as **projected** from the side elevation to that of the end elevation by the use of the imaginary horizontal **projection lines** that start at the opening as drawn in the side elevation through the end elevation.

The lines named with capital letters that pass through the end elevation of Fig. 110 are shown in the end of the Die Block "M" of Fig. 109 with small letters, should enable you to compare them with those lettered with capital letters.

In Fig. 109, the cross hatched end of the Die Block "M" is all that is seen in the end elevation of Fig. 110. The side "N" of the Die Block is not to be thought of when looking for information in the end elevation. In order to see any of the "N" half of the Die Block, it is necessary to show another cross sectioned end elevation on the other side of the side elevation, which would have to be a L-H. end elevation, which is not needed in Fig. 110. The end elevation as drawn answers all that is necessary. Any other information that may be necessary can be gleaned from the side and end elevation of Fig. 110, is easily understood by referring to the separation of the Die Block of Fig. 109.

In Fig. 113 is shown a blue print drawing of an Engine Cylinder. Figure 111 and 112 help show another example of a **zig zag cross section** blue print drawing.

It will be noticed in Figs. 111 and 112 that only half of the Cylinder in picture-like form are needed in showing the elevations and plans of Fig. 113.

The three plan cross sections shown in the top of the blue print drawing of Fig. 113 are to be imagined as taken from different positions noted along the **zig zag lines** shown in the side and end elevations.

All of the three plan cross sections shown in the top of the blue print drawing of Fig. 113 could have been placed anywhere on the blue print drawing of Fig. 113 as long as the notations as seen placed below these cross sections, were also placed below the same sections wherever they might have been placed on the blue print drawing of Fig. 113 so as to mark them in relation to the side and end elevations.

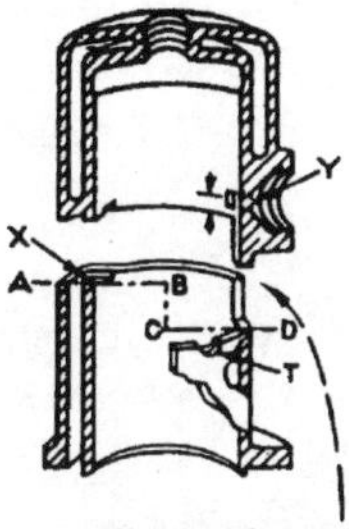

Fig. 111

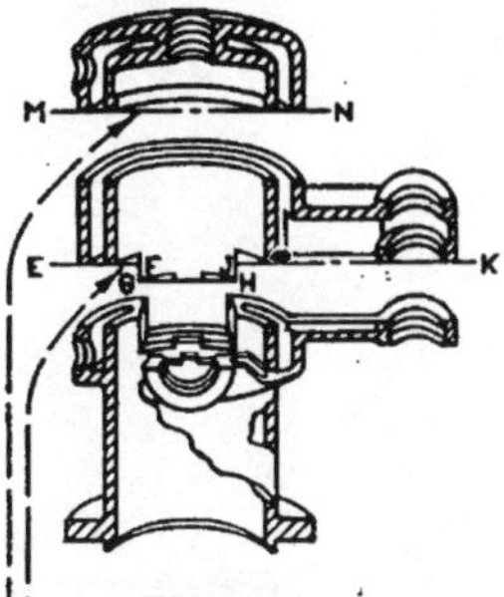

Fig. 112

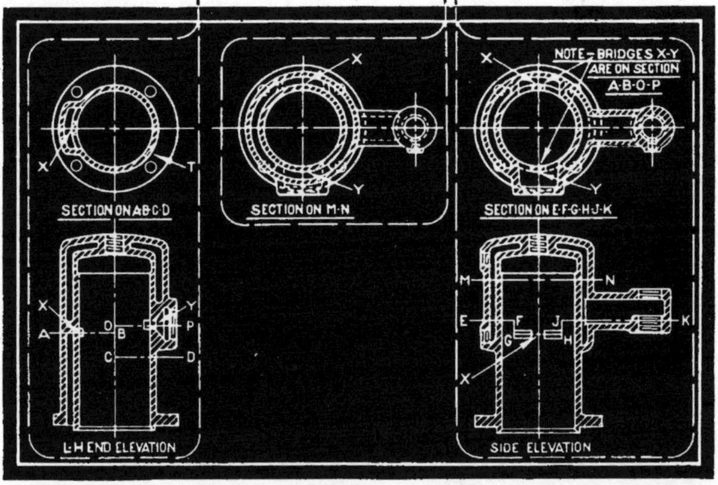

Fig. 113

The meaning of the line "M-N" in the "SECTION ON M-N" shown in the side elevation of Fig. 113 is understood by examining the plan cross section shown in the center and at the top of the blue print drawing.

The encased broken arrow line which directs to the top of Fig. 112 gives an easier way of understanding the meaning of the line "M-N" of the side elevation of Fig. 113.

The line "M-N" of the side elevation of Fig. 113 is a straight line cross section, which is like almost all common cross sections. It is unlike the other two cross sections shown in the blue print drawing of Fig. 113.

The plan cross section "E-F-G-H-J-K" as seen above the side elevation of the blue print drawing of Fig. 113 is not the same in principle as the cross section "M-N," because the cross section "M-N" was taken on a straight line, and the cross section "E-F-G-H-J-K" is taken from the **zig zag line** "E-F-G-H-J-K" of the side elevation of Fig. 113. By means of the broken arrow line pointing to Fig. 112, the **zig zag cross section** "E-F-G-H-J-K" may be easily understood.

The **zig zag plan cross section** "E-F-G-H-J-K" as shown above the side elevation of the blue print drawing of Fig. 113 may also serve the purpose of showing the bridges "X" and "Y" from two separate planes to be on one plane, thus showing the necessary detail in one cross section without making another cross section.

Should there be no note above the cross section plan "E-F-G-H-J-K" concerning the bridges "X" and "Y" stating that they were not on the **zig zag plane,** then the section "E-F-G-H-J-K" as shown above the side elevation of Fig. 113 would be understood as on the **zig zag plane** of the cross section plan.

Without the note shown in the cross section plan "E-F-G-H-J-K" concerning the location of the bridges "X" and "Y," another cross section plan would be necessary.

Should a straight line cross section be taken from the side elevation from "E" to "K" instead of from the **zig zag line** "E-F-G-H-J-K," what would be seen in the top plan cross section of the bridge "X" and "Y" would be drawn with dotted (hidden) lines. A **zig zag cross section** can therefore be more clearly understood, because it lays practically everything in one plane, open to full view.

The section line "A-B-O-P" as shown in the L-H. end elevation can be used to show by the dimension arrows, how far apart these openings are. See Fig. 111.

The cross section plan as placed above the end elevation cross section of the blue print drawing of Fig. 113, is taken from the **zig zag line** "A-B-C-D," which is given in Fig. 111.

The object of the **zig zag cross section** "A-B-C-D," is to show the bridge "X" and the thread "T," which are on two planes of Fig. 111, as if shown on one plane in the cross section plan above the end elevation of Fig. 113, in full view of one cross section without the use of any dotted (hidden) lines.

RIB CROSS SECTION

In Fig. 116 is shown a blue print drawing of a Bracket with two elevations and a top plan to its side elevation.

As stated below the L-H end elevation, "SECTION ON C-D," is to be thought of as a cross section taken through the center line "C-D" of the top plan. The object of this problem is not for the purpose of showing how a cross section is to be understood in relation to its other elevations, for that has been previously explained, but for the purpose of making understood the particular cross hatching shown.

In the L-H. end elevation of Fig. 116 is shown the main body of the Bracket detail with close-spaced line cross hatching. Its rib is spaced twice the distance of the spacing of the main body of the Bracket, or in other words, the rib cross hatched lines are every other line from the cross hatched lines of the main body of the Bracket detail.

The object in cross hatching the rib every other line from the main body of the Bracket, is that a rib of any article that runs **length ways** may show by the cross hatching that it does belong to the main body proper, but as a solid connecting member only.

The cross section of Fig. 114 shown by the dotted (hidden) lines "E-F-G" suggests the difference in thickness between the main body of the Bracket and that of the rib.

Should a cross section be taken from the line "A-B" of the R-H end elevation of Fig. 116, the rib would be considered in the R-H. end elevation as a solid connecting member to the main body of the

Bracket as it is a regular cross section, because it was cross sectioned **crosswise.**

A rib of any main body, as shown in the L-H. end elevation of Fig. 116, would show cross hatching every other line apart from the main body, because it was cross sectioned **lengthwise.**

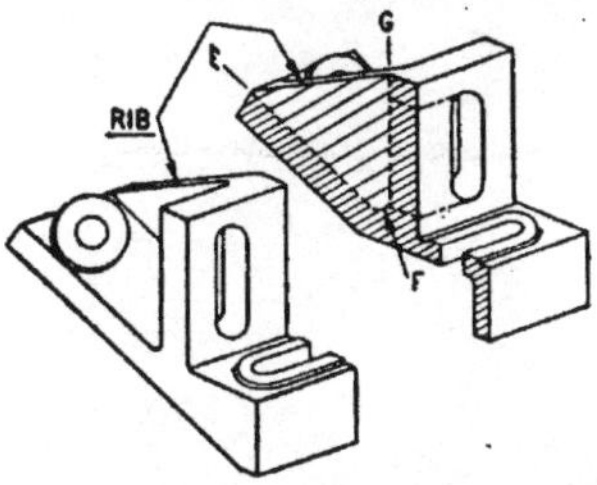

Fig. 114

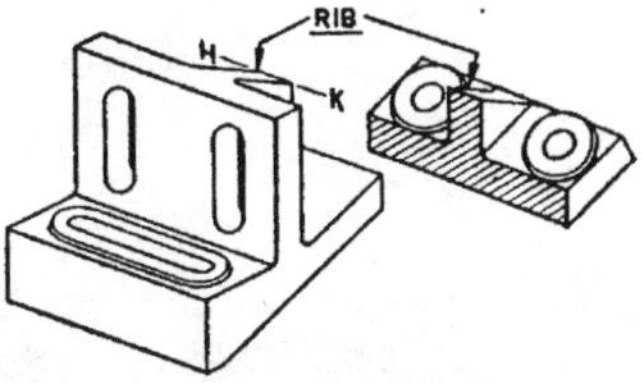

Fig. 115

Care and study of the above mentioned rules should be a watchword in all blue print drawings that show ribs and thin surfaces.

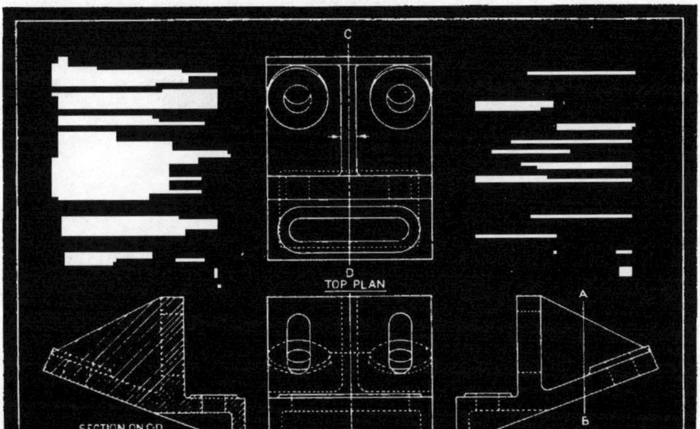

Fig. 116

There is to quite an extent, a big practice among draftsmen not to cross hatch a rib or thin surface as explained with Fig. 114, and such unexperienced practice of draftsmen should likewise be taken note of.

CREATED LINE CROSS SECTIONS

Figures 118 and 120 are each a separate and complete blue print drawing of Figs. 117 and 119 respectively.

The R-H end elevation of Fig. 118 shows the posi-

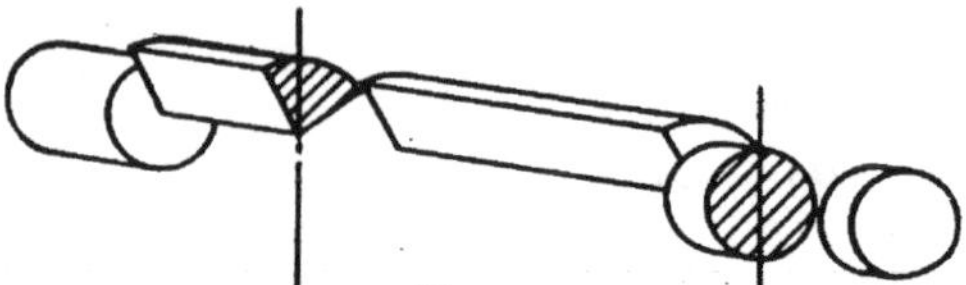

Fig. 117

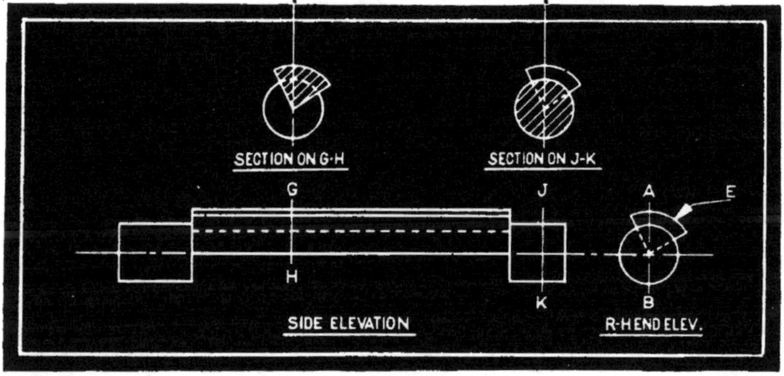

Fig. 118

tion of the part "E" in relation to the center line "A-B," which end elevation is all that is needed to convey the meaning of the blue print drawing of Fig. 117.

Should a cross section be desired in Fig. 118 from either of the **created lines** "G-H" or "J-K," such cross sections at these particular points could be placed anywhere on the blue print drawing, so long as it is stated below these cross sections where they were taken from, however, with the R-H. end elevation as shown, none of the cross sections need be shown for clear understanding.

The cross sections "G-H" and "J-K" as shown in Fig. 118, were not taken from any common center line, because there is no vertical center line in the side elevation, as those cross sections were taken from **created center lines** that was desired in showing cross sections at these particular points.

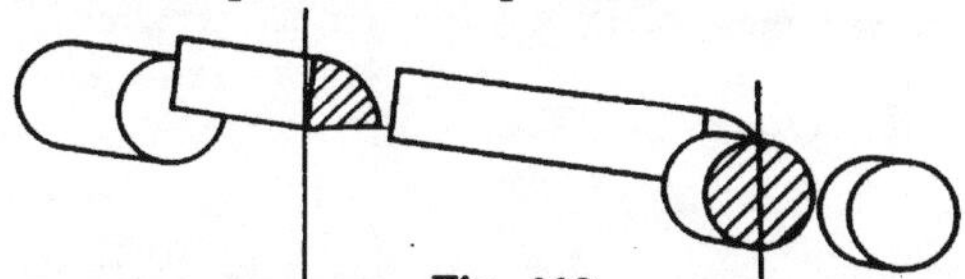

Fig. 119

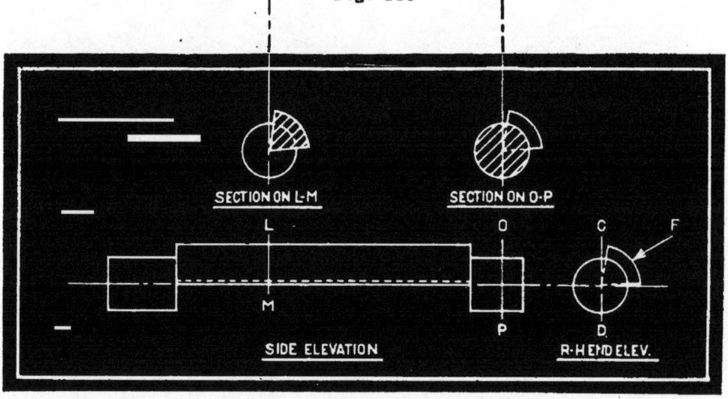

Fig. 120

Either one of the cross sections "G-H" or "J-K" could have been placed where the R-H end elevation is now placed in Fig. 118, but in either case, the place where that cross sectioned end elevation came from, would necessarily be stated below the cross section of where it came from.

The vertical center line of the R-H end elevation of Fig. 118 did not need the "A" and "B" placed on each end as is shown. These were only so used to call attention to the part "E" sets in relation to that center line "A-B," and likewise the part "F" in relation to the center line "C-D" of the blue print drawing of Fig. 120.

What has been explained of Figs. 117 and 118 applies to Figs. 119 and 120 as well.

COMPLICATED CROSS SECTIONS

Complicated cross sections are not different in meaning from the ordinary cross section, only that it is confusing in showing a blue print drawing of a particular make up of an article, such as is shown in the blue print drawing of Figs. 123 and 125.

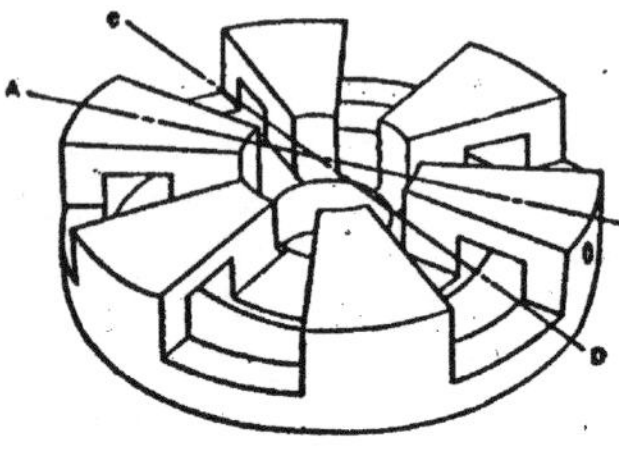

Fig. 121

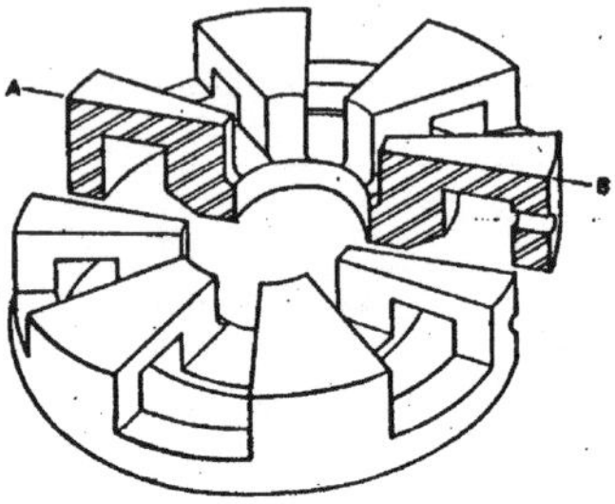

Fig. 122

A **complicated** R-H end elevation is shown in the complete blue print drawing of Fig. 123 that was made of the object shown in Fig. 121.

The end elevation of Fig. 123 is a cross section, which is taken from the center line "A-B" of the side elevation. This center line is the same center line "A-B" of Fig. 121, as given in the cross sectioned end of Fig. 122.

The top plan of Fig. 123, shows the outside top end shape of the side elevation.

The value of the cross sectioned end elevation of Fig. 123 can be grasped from the perspective drawing of Fig. 122. If the end elevation of this were not cross sectioned, it would look just like the top plan of Fig. 123, and likewise, would show the exceptional amount of dotted (hidden) lines in a confusing way.

Many blue print drawings are shown with an end elevation of a complicated object, with no cross section, which can be understood with projection lines, but such a blue print drawing as that of Fig. 123 would most likely have a cross sectioned end elevation to help an understanding, like that shown of the end elevation of Fig. 123.

Figure 125 is likewise a complete blue print drawing of Fig. 124, but taken from the center line "C-D"

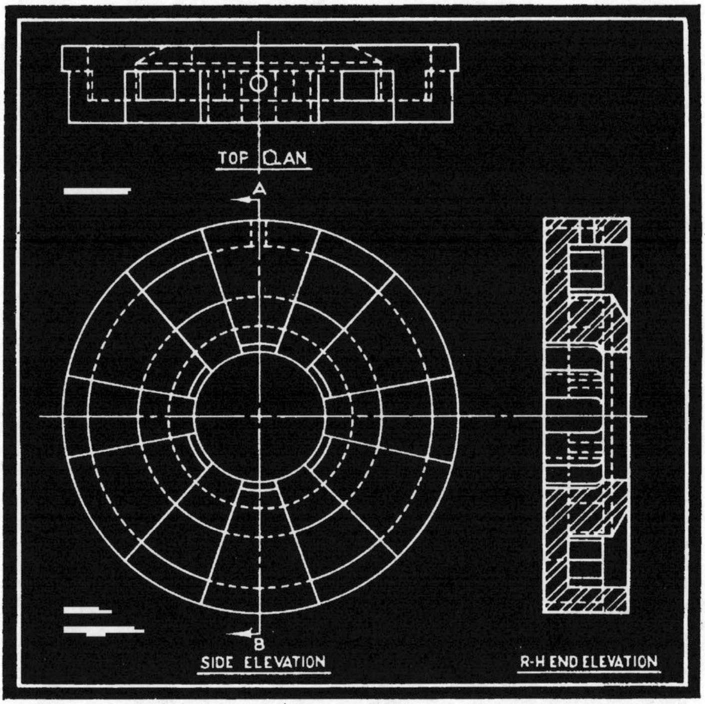

Fig. 123

instead of the line "A-B" of Fig. 121. The center line "C-D" of the object is shown in Fig. 124.

The end elevation of Fig. 125 is also much more easily understood than its top plan, because it is a cross sectioned end elevation; for the principle of a cross section is to show an end elevation drawn to represent the center of an object imagined to be cut into

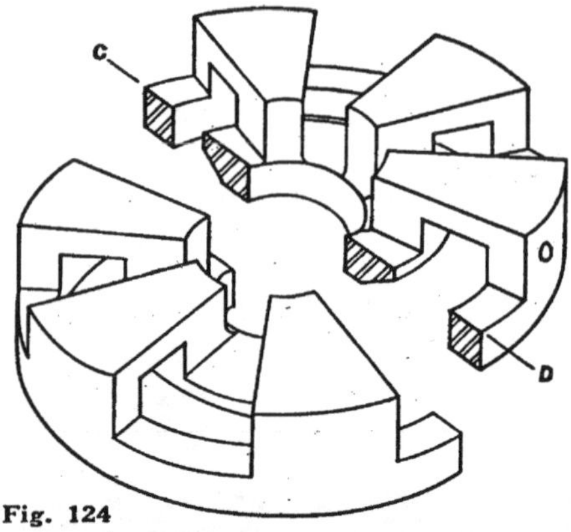

Fig. 124

two parts, so that the center can be shown on a cross section without the outside being in front of it.

If the end elevation of Fig. 125 were not cross sectioned, then inside of the elevation would be drawn with dotted lines, like those in the top plan.

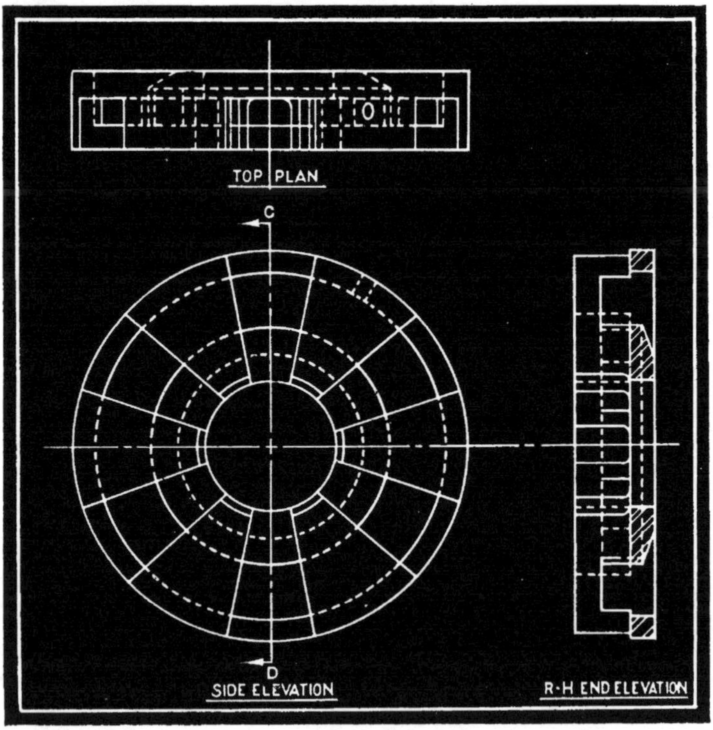

Fig. 125

THE WORDS "ELEVATION OR PLAN" ELIMINATED FROM BLUE PRINT DRAWINGS

From the beginning of these instructions the words **"views," "elevations"** or **"plans"** were shown on the different blue print drawings as a guide to the reader with the explanations of the detail.

While the words **"elevation"** or **"plan"** as used were correct, these words are seldom ever shown on any blue print drawing, as the reader of a blue print drawing should know when looking at an **elevation** or **plan** that they are such.

From now on, no more such words as **"elevation"** or **"plan"** will be shown in the blue print drawing, such positions should be fully located by the reader.

PART VII

ASSEMBLY BLUE PRINT DRAWINGS

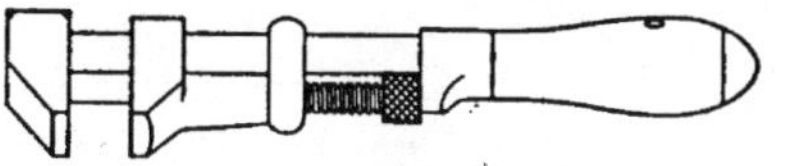

Fig. 126

An **assembly blue print drawing** implies from the word **"assembly"—place together.**

As practically all objects that are to be made for any purpose consist of several parts, each of these separate parts when drawn separately are called **"details,"** which was dwelt on in the first part of this book. When these parts are all placed together, they form an **assembly blue print drawing** as Fig. 127 shows.

Figure 126 is a perspective drawing of a Monkey Wrench and Fig. 127 as its **assembly blue print drawing.**

An **assembly blue print drawing** serves the purpose of showing one part's position in relation to that of another part, besides showing such parts as are openly seen, so that these can be judged for their importance as to fit, finish, etc. The **assembly blue print drawing** serves chiefly to show where each piece is to be placed in the **"assembly."**

The Monkey Wrench of Fig. 126 is a tool that every man has handled, and which is easily understood. Should the blue print drawing of Fig. 127 be

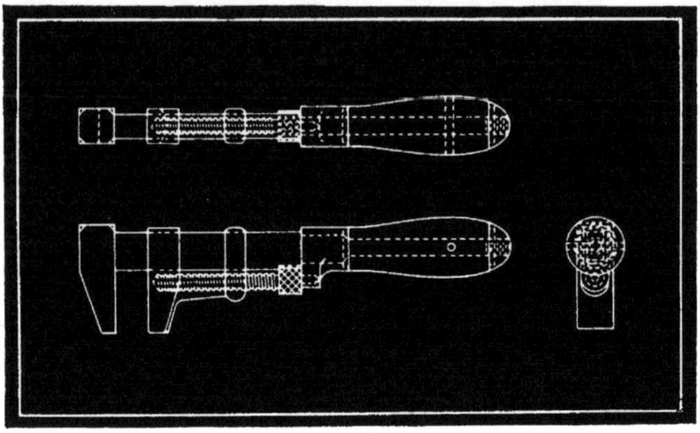

Fig. 127

not readily understood, secure for comparison, a Monkey Wrench like the one shown.

Figure 166 shows all of the parts composing the **assembly blue print drawing** of Fig. 127, drawn in a group, and explained under the chapter entitled "GROUP DETAIL BLUE PRINT DRAWINGS" that will be treated in the next few pages.

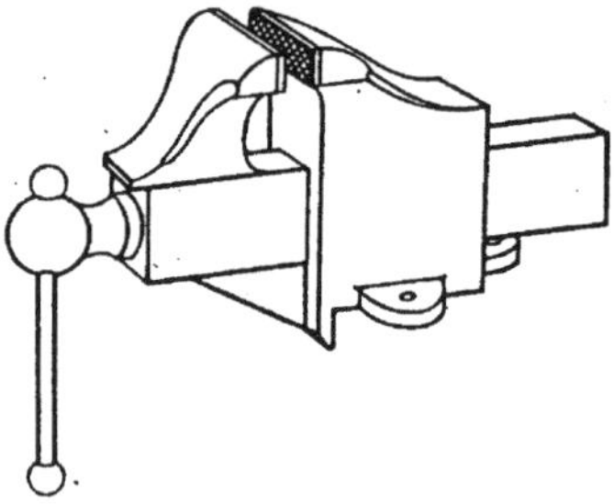

Fig. 128

In Fig. 128 is shown a perspective drawing of a Machinist's Vise that you no doubt are familiar with. Fig. 129 is its **assembly blue print drawing.**

Each piece of the Machinist's Vise as composed in the **assembly blue print drawing** of Fig. 129, is drawn with separate elevations and a top plan.

Each part that it takes to make up the Vise is shown in the group blue print drawing of details, as shown in Fig. 130.

When each separate part detail in the group of Fig. 130 is made up, it should then be **assembled,** or in other words, **"put together"** from the guidance given in the **assembly blue print drawing** of Fig. 129.

The complete group blue print drawing of details of Fig. 130, conveys all the information that is neces-

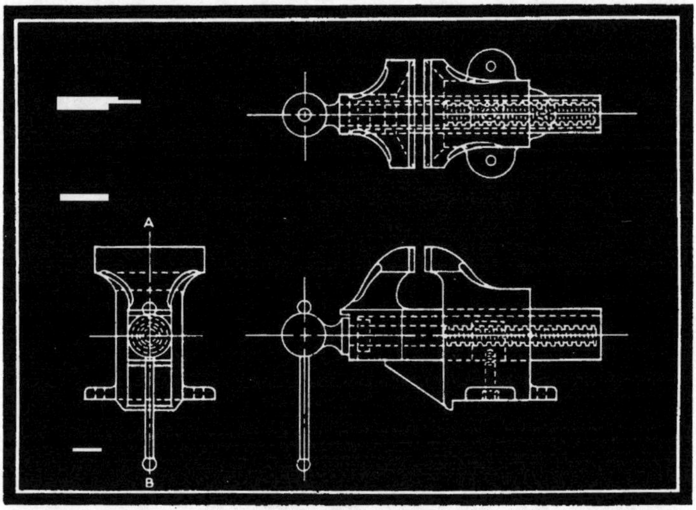

Fig. 129

sary for a mechanic in making each part for the Vise shown in the **assembly blue print drawing** of Fig. 129, should dimensions be given on each detail.

In the side elevation of the **assembly blue print drawing** of Fig. 129, what is seen of the inside of the vise is shown with dotted lines, because the draftsman saw that a cross section was unnecessary to show that detail.

Should the draftsman feel that the inside of the Vise did not give as full meaning with the dotted lines

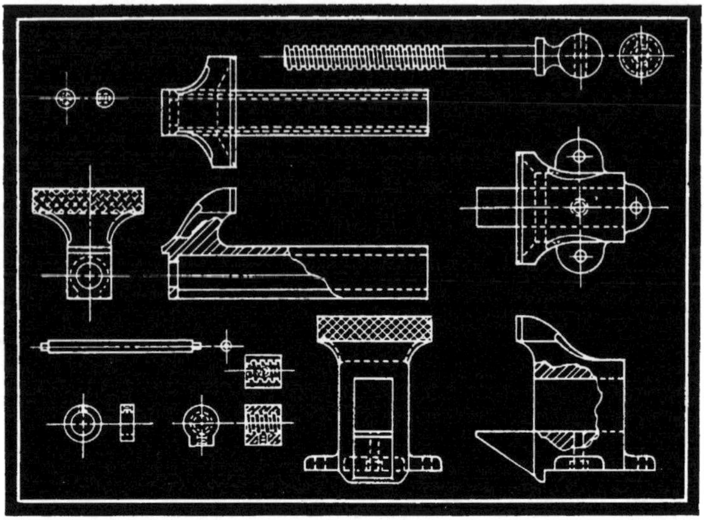

Fig. 130

as it would if the side elevation was a cross section, (where the use of full lines would be shown inside of the Vise), he would perhaps show a cross section in the side elevation of Fig. 129 like that of Fig. 131, which is given fully in Fig. 132.

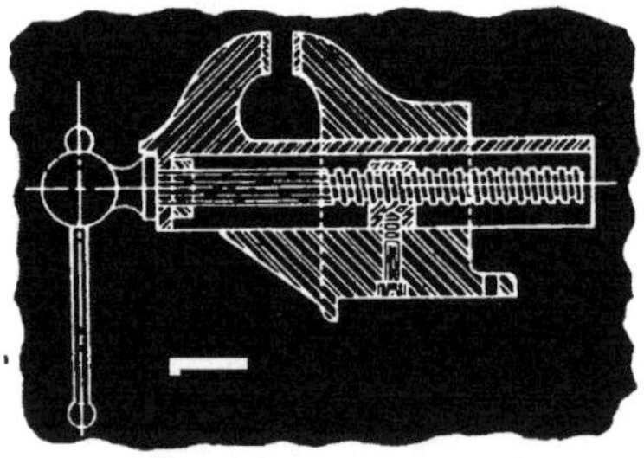

Fig. 131

Should Fig. 131 be applied to the side elevation of Fig. 129, it would be shown as a center line cross section of the side elevation of the Vise as taken from the center line "A-B" of the L-H. end elevation.

In Fig. 133 is shown the Jaw Body part of the Vise drawn with two elevations and a top plan, as taken from the group of details of Fig. 130, for a blue print drawing detail.

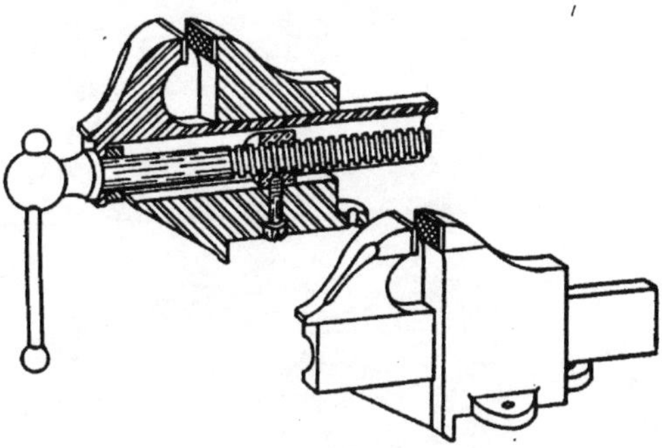

Fig. 132

In the side elevation of Fig. 133 is shown a broken out cross section, which is to show a part of the opening "C" of the L-H. end elevation that the Sliding Jaw is to go into, so that the opening "C" of the L-H. end elevation can be partly seen in the side elevation with-

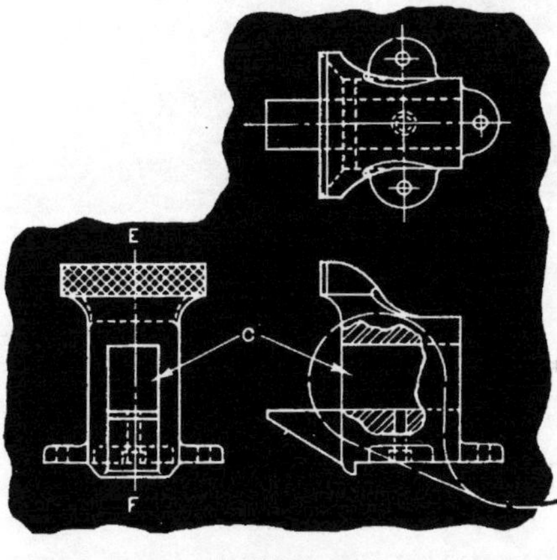

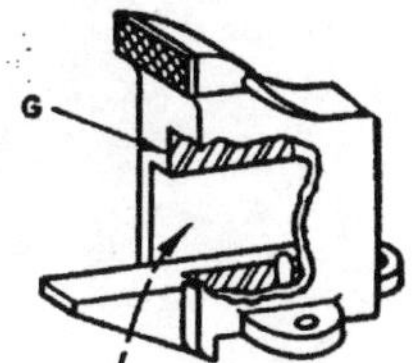

Fig. 134

Fig. 133

out using altogether dotted (hidden) lines to show where that opening is.

The opening "C" as shown in the side elevation of Fig. 133 also shows a part of the hole that is in the bottom part of the Body Jaw, open. One half of this bottom hole is covered by the outside lower part of the Jaw Body, and to get an idea how the broken out cross section of the side elevation of Fig. 133 really looks, follow the dashed arrow line that points in the perspective drawing of the Jaw Body of the Vise in Fig. 134.

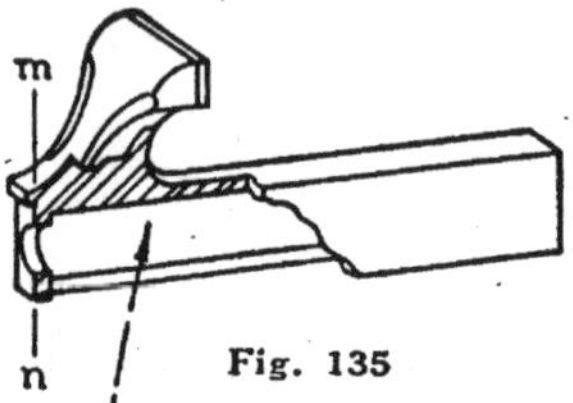

Fig. 135

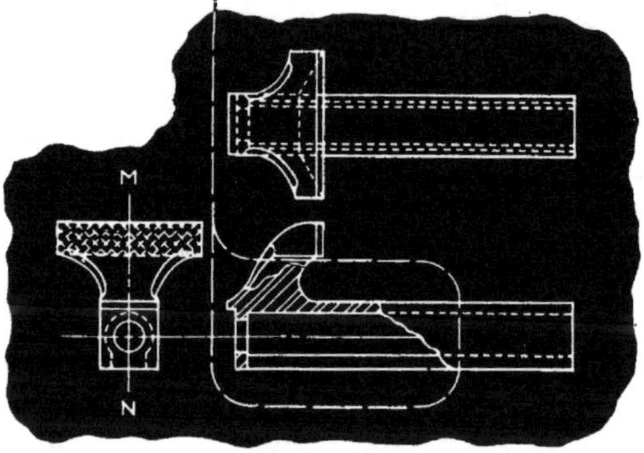

Fig. 136

What is shown cross hatched in the broken out section of the side elevation of Fig. 133, is on the center line "E-F" of the end elevation where the cross section is taken from. The center line "E-F" of Fig. 133 would be the line "G" in Fig. 134.

In the detail of the Sliding Jaw that was shown in Fig. 130, and shown separately in Fig. 136; the side elevation offers another example of a broken out cross section for a blue print drawing. The broken out cross section of Fig. 136 is taken from the center line "M-N" of the end elevation, which can be clearly understood by studying the dashed arrow line in the direction of the perspective drawing of the Sliding Jaw of Fig. 135 center line "m-n."

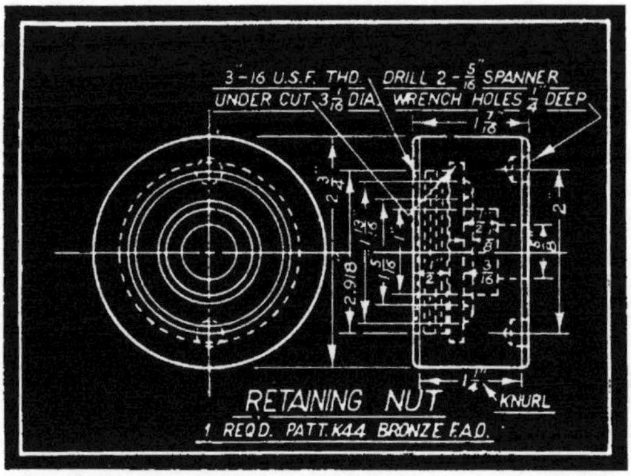

These two blue print drawings are for the same object, the bottom one shows a (broken out) cross section; the top one without. Which is the easiest one to understand?

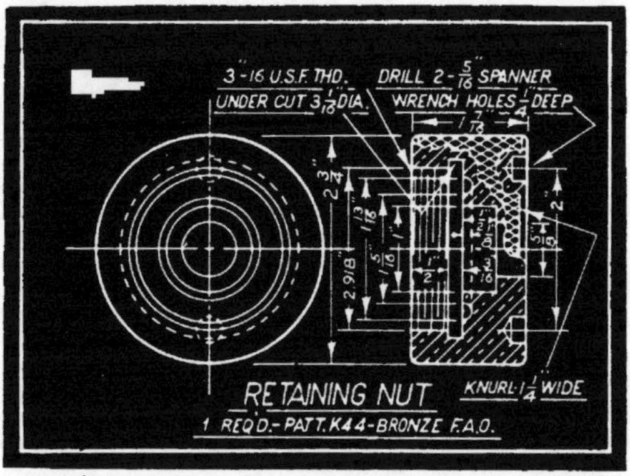

Practice Reading Figures

ENLARGED ASSEMBLY BLUE PRINT DRAWING

It is often the case in an **assembly blue print drawing,** that an actual size drawing would be so small, that the necessary detail could hardly be seen.

Figure 139 would perhaps be the actual size that a blue print drawing of a Lever was to give detail to, but the draftsman would perhaps make his blue print drawing detail the size shown in Fig. 140 many times larger, so that all of the necessary detail could be clearly seen.

Figure 140 is not drawn to full height, as Figs. 138 and 139 are. This is shown by the broken left out section. Any further detail concerning this broken left out section would necessarily be noted in its blue print drawing detail.

In Fig. 140, every detail of the Lever as shown can be seen to good advantage, while such details can hardly be seen in Fig. 139 on account of being so small.

When an elevation, as in Fig. 140, is made larger than the actual size detail shows, it should have a remark, such as, "ENLARGED ELEVATION," placed below the enlarged elevation as drawn, so as to show that what is drawn is more than the actual size.

If a blue print drawing like that of Fig. 140 be desired shown to a Scale, then instead of such a remark being placed below the elevation as "ENLARGED ELEVATION," the remark would then state how many times larger than actual size the elevation is, such as "DOUBLE SIZE,"—"SCALE—6″=4″," or whatever the extra size happens to be.

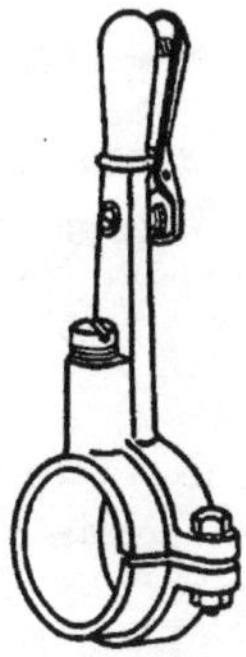

Fig. 137

Fig. 138

Fig. 139

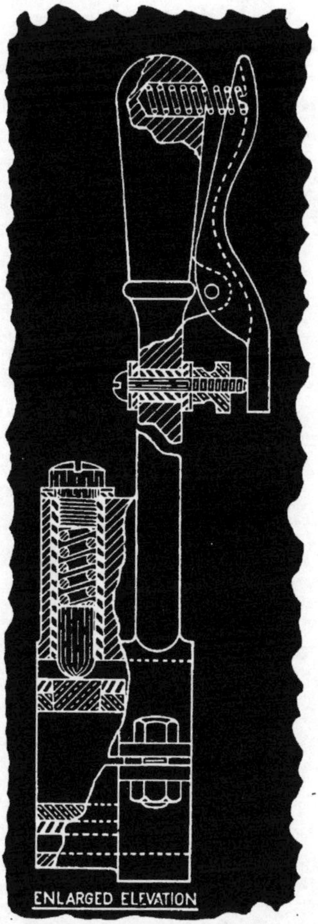

Fig. 140

ASSEMBLY CROSS SECTIONS

In Fig. 142 is shown an assembly blue print drawing of a Slide with Figs. 141, 143 and 144 to help illustrate the explanations, and convey to your mind, how two or more pieces when drawn into a complete assembly blue print drawing are to be understood when an assembly shows a cross section.

In the end elevation of Fig. 142 is shown a cross section as taken from the created line "A-B" of the side elevation which is noted as such below the cross sectioned end elevation.

When a remark is given below the cross section of a blue print drawing, one is to look somewhere in the next elevation for a line that the cross section was taken from, as for example, "SECTION ON A-B" in Fig. 142. In this case, the side elevation has no common center line that a cross sectioned end elevation could be obtained from, so by placing below the end elevation of Fig. 142, such remark as "SECTION ON A-B," it is stated where the cross section was taken from.

The line "A-B" of the side elevation of Fig. 142 as just explained is not considered a natural center line of the side elevation, because the center lines are generally considered to be the center of a round object, but should the line "A-B" be the center line of the side elevation of Fig. 142, and the cross section as shown in the end elevation be taken from this **supposed** center line "A-B," then the remark "SECTION ON A-B" is not needed below the cross sectioned end elevation of Fig. 142, as the cross section end elevation could then be obtained from the center line "A-B" of the side elevation. For example, the cut-in-two section of Fig. 141 shows the supposed center line "A-B," which is really a created line of the side elevation that the cross section is taken on.

In Fig. 142, the blue print drawing is shown complete with side and cross sectioned end elevation with its top plan, which shows the Slide with its two Gibs, and the Slide Body which are all placed together as shown in Fig. 141.

The cross sectioned end elevation of Fig. 142 is

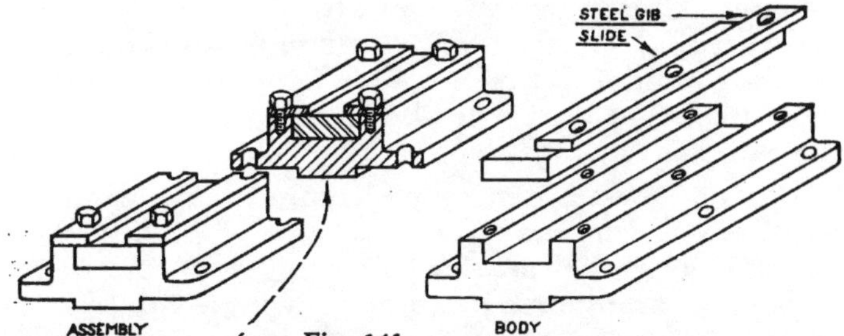

Fig. 141

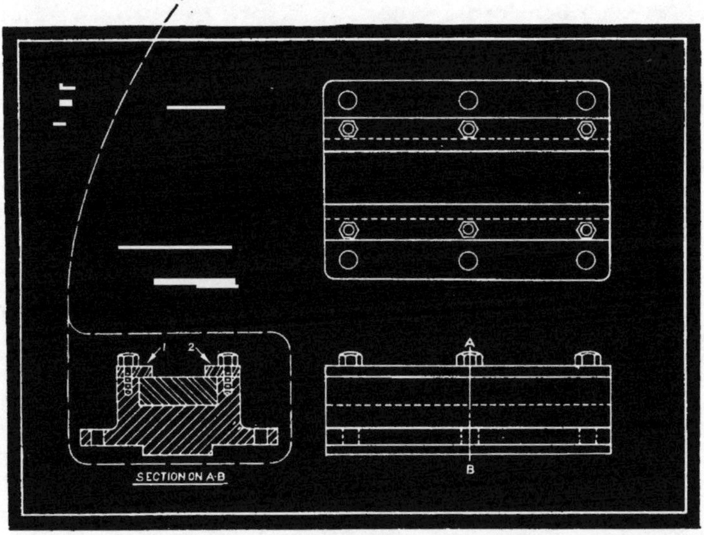

Fig. 142

taken from the created line "A-B" of the side elevation. It may be more clearly understood from the dashed arrow line that encases the end elevatiton of Fig. 142 pointing to the cut-in-two section in Fig. 141.

The Body part, besides that of the two Gibs and the Slide are cut-in-two, as all cross sections mean, like the cross section of the end elevation of Fig. 142.

Cross sections are always known by the presence of cross hatchings, so one can always understand that cross hatching means that the elevation or plan that is a cross section is taken somewhere from the elevation or plan that is next to the elevation or plan that is cross hatched.

To one who always keeps in mind the meaning of the different symbols of materials as Fig. 89 shows, which has a general meaning to every one who understands the reading of blue print drawings, would know at a glance what kind of material is being used in the different parts cross hatched. Note the cross section end elevation shown in Fig. 142.

There are four separate parts besides the Screws shown in the cross sectioned end elevation of Fig. 142, which are all to be understood by cross hatchings. These either have a different meaning given by the material symbols, or the lines of the cross hatchings for each part slant in different direction or angle from the other, so as to separate the parts from one another.

The Body part, according to the end elevation of Fig. 142 is to be cast iron, as shown by the cast iron symbol of cross hatching; the Slide and the Gibs are to be in steel, as shown by the spacing and design, all different from that of cast iron. Each symbol showing a part slanting in an opposite direction from the part that is next to it, is to know the separate parts composing the **assembly.** In this way, an **assembly** is understood for its separate parts by whatever symbol or angles the cross hatching shows for each part.

In Fig. 143 is shown an end elevation almost like that of the end elevation of Fig. 142, only that the Gib is drawn in the cast iron symbol of material, instead of steel as was shown for the Gibs of the end elevation of Fig. 142. The Gib of Fig. 143 is made in one piece, rather than in two pieces as drawn in Fig. 142.

The purpose of showing the Gib of the end elevation of Fig. 142 in the symbol of steel, and the Gib of the end elevation of Fig. 143 in the symbol of cast iron, is to show how the cast iron symbol cross hatching for the Gib of Fig. 143 would be cross hatched to slant in one direction, while the cast iron symbol of Material of the Body part is cross hatched to slant in the opposite direction so that the Gib may be known from that of the Body as being the same material, but separate parts.

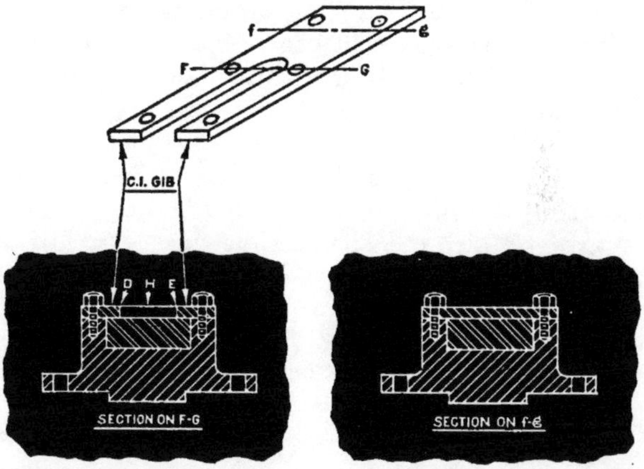

Fig. 143 Fig. 144

To know in Fig. 142 that the Gib No. 1 is not a part of Gib No. 2, is to be able to understand when looking at a cross section, that whenever cross hatched lines of the same symbol slant in the same direction and at the same angle, and are apart from one another as shown of the Gibs of Fig. 142; they are one and the same piece connected elsewhere in back of the center line that the cross section is drawn. It is not always to be figured that they are one, only that this is very likely. In order to know, you are to look carefully at every elevation and plan of the blue print drawing of Fig. 142 for a possible connection of Gib No. 1, to that of Gib No. 2.

In both the end elevation and top plan of the blue print drawing of Fig. 142, there is no connection of Gib No. 1 with the Gib No. 2, which you will notice if you study the top plan, and the side and the end elevation.

Should the end elevation of Fig. 143 be in place of the end elevation of Fig. 142, and the slant lines of the cross hatched symbol in the same direction on each end of the Gib as shown in Fig. 143, the Gib would be made in one piece by the addition of line which runs from "D" to "E."

The Gib as drawn in the end elevation of Fig. 143 by connecting the line from "D" to "E" to make it one piece, might be like the shape that the perspective cast iron Gib shows in Fig. 143. The line "D-E" would be what is shown connecting one side of the Gib to the other side of the Gib in back of the line "F-G" of the perspective cast iron Gib of Fig. 143.

Should the cross section of the Gib of Fig. 143 be taken as far back as the broken line "f-g," instead of through the line "F-G," of the perspective cast iron Gib, the opening part "H" of the Gib would be cross hatched in continuation with the same symbol that the ends of the Gibs now show, as in Fig. 144, and there would be no vertical lines that lead from "D" and "E" of Fig. 143, as is now seen in the opening "H" of the Gib of Fig. 143.

Should the perspective Gib of Fig. 143 be the Gib for Fig. 142, the line "A-B" of the side elevation would then be the line "F-G" of the perspective Gib of Fig. 143, and the end elevation of Fig. 143 would be in place of the end elevation of Fig. 142.

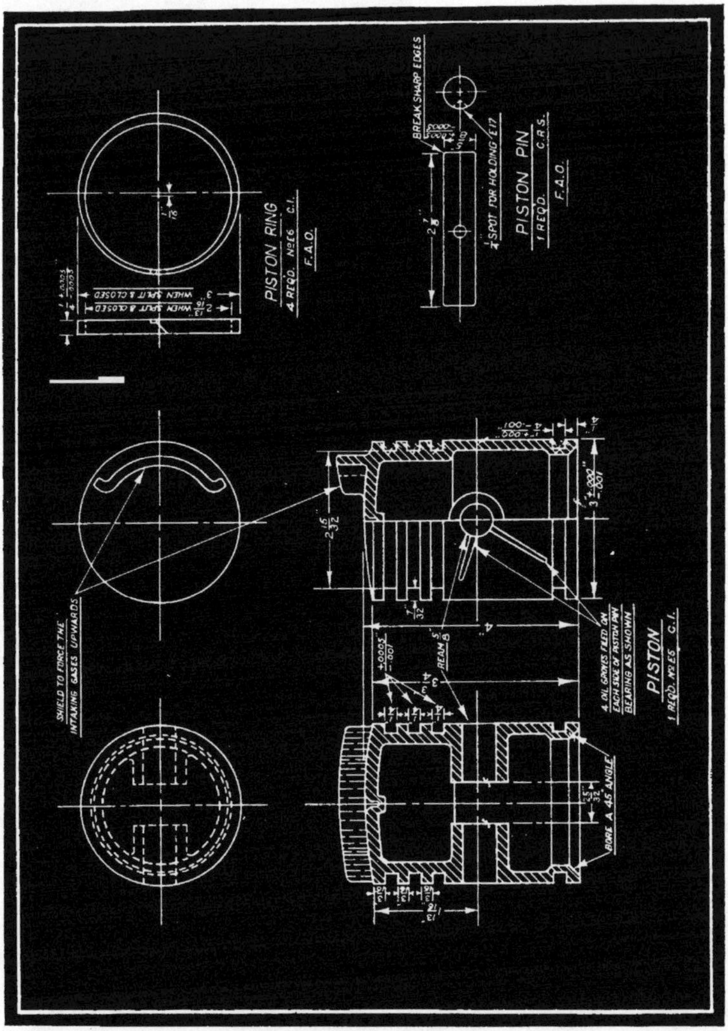

Practice Reading Figure

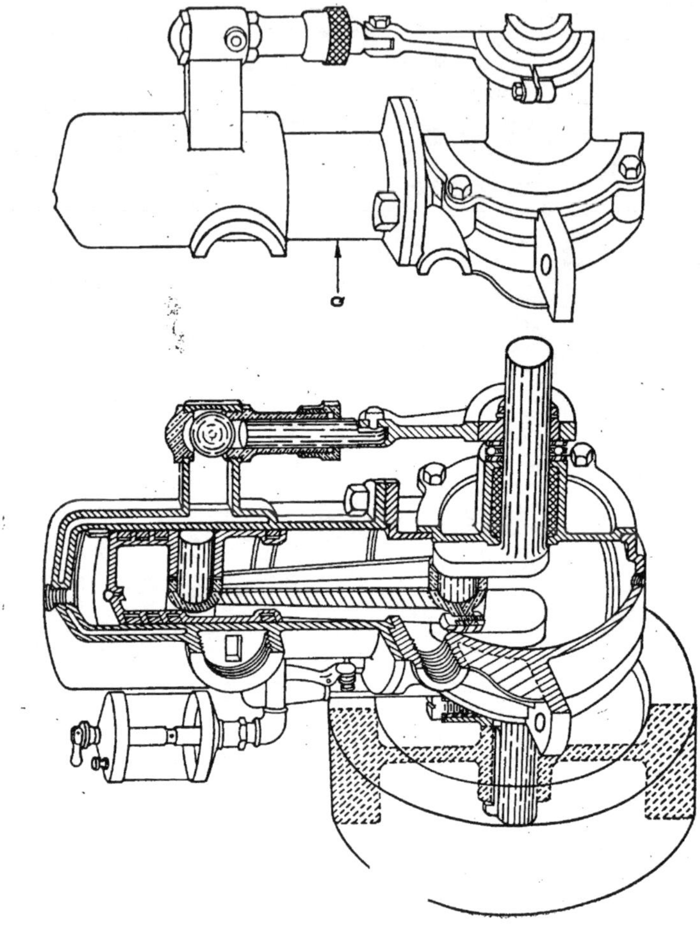

Fig. 145

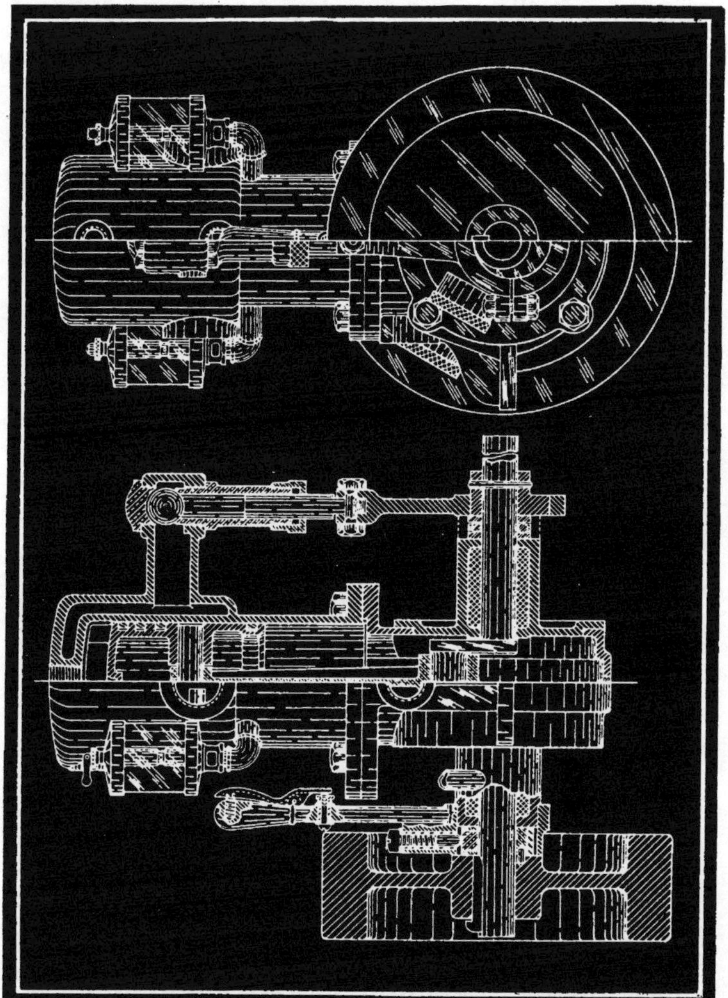

Fig. 146

In Fig. 146 is shown an assembly blue print drawing of an Engine with a side and an end elevation.

In the end elevation of Fig. 146 are contained a half of the **front** part, with a half of the **back** part of the Engine drawn butted together on the center line.

The object of a draftsman in drawing the end elevation with two halves, is to give detail of the **front** and **back** part of the Engine, so as to convey just

Fig. 147

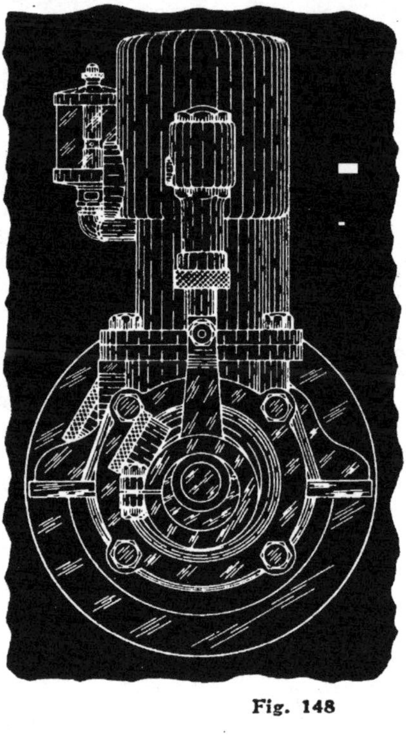

Fig. 148

enough information of all that is necessary for an understanding of an assembly blue print drawing of Fig. 146 without going to unnecessary work in drawing a separate end elevation for the **front,** and also another separate end elevation for the **back part** of the Engine.

As has just been explained, that a half of the **back** end elevation of Fig. 146 is the half of Fig. 147 that Fig. 148 shows, and the half of the **front** end elevation is the half of Fig. 150 that Fig. 149 shows. These halves are joined together at the center line in the end elevation of Fig. 146, so as to show a reasonable amount of the **front** and **back** part of the Engine in one elevation for you to imagine would look. They would

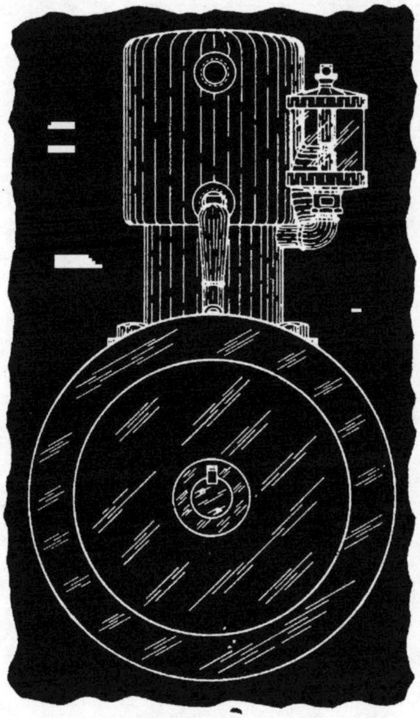

Fig. 149

Fig. 150

look in reality as in Fig. 148 does for Fig. 147, and Fig. 149 for Fig. 150.

A mechanic who is not familiar with the fundamental principles of reading a blue print drawing, would no doubt take his meaning from the end elevation of Fig. 146, by getting the idea that the Engine is drawn fitted up with two Oilers, which is not so.

By looking carefully at the **back end** of the Engine of Fig. 148, one can see only one Oiler in place, which is on the left hand side, and when looking at the **front end** of the Engine of Fig. 149, that Oiler is on the right hand side, because each elevation of each **end** of the Engine is seen from the opposite direction; so that when the **front** and **back end elevations** that represent the halves of Figs. 147 and 150 are butted together on the center line to form the end elevation of Fig. 146, they are apparently two Oilers, although as just explained, this is not the case.

The assembly blue print drawing of the Engine of Fig. 146, besides what has thus far been explained, conveys information for an assembly blue print drawing without the use of any other elevations or plans, for an assembly blue print drawing as given to the mechanic is practically to be used only to show the general position of the parts, when the several parts as composed in the detail blue print drawings are placed together.

In order to understand the assembly blue print drawing of Fig. 146 more fully, look at the top plan of Fig. 151 as if the same were placed over the side elevation and included in the whole blue print drawing of Fig. 146.

The full lines as drawn in the top plan of Fig. 151 are to show that part of the side elevation that is given in Fig. 146. The **section "Q"** that is drawn with dashed lines, is that part that was left out to show the cross section as seen in the side elevation of Fig. 146.

The **section "B"** of Fig. 151 is what was taken out of the side elevation of Fig. 146, so as to show more of the Timer Lever that projects into the Wheel. Fig. 146 also shows the cross section shape of the Wheel in the assembly blue print drawing without the mechanics having to look at a separate detail blue print drawing of the Wheel for its cross section.

What has been explained of the cross section of the Engine thus far, can be more simply understood by referring to Fig. 145, which clearly sets out for an understanding of what the **quarter section** "Q" of Fig. 145 shows, as was left out to make the cross section as shown of the side elevation of Fig. 146.

What is shown of the perspective drawing of Fig. 145 is the impression the mechanic should imagine of

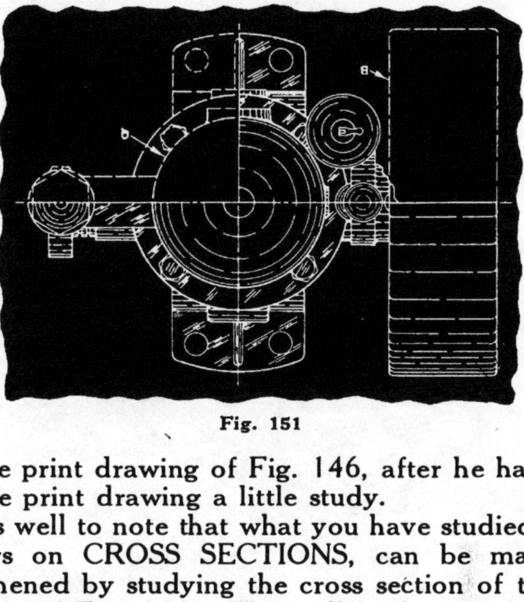

Fig. 151

the blue print drawing of Fig. 146, after he has given the blue print drawing a little study.

It is well to note that what you have studied in the chapters on CROSS SECTIONS, can be materially strengthened by studying the cross section of the side elevation of Fig. 146. This will be known from the top plan of Fig. 151, as that which is shown drawn with full lines. It shows that the cross section in the side elevation of Fig. 146 is a quarter cross section, because it belongs on two center lines.

In Fig. 153 is shown another assembly blue print drawing on an Engine that has all of the **back end** only in the end elevation as one elevation. This is not like the half elevations as shown in the end elevation of the blue print drawing of Fig. 146.

The end elevation of Fig. 153 has more to show of the **back** than could be shown to any advantage for the **front** part of the Engine.

The cross section as shown in Fig. 153 is worthy of careful consideration and study, so as materially to strengthen understanding of an assembly blue print drawing.

In Fig. 153, the **assembly blue print drawing** shows only a side and end elevation. To understand the cross sections shown for Fig. 153, imagine a top plan placed above, as if taken from the side elevation of Fig. 153 and like that of Fig. 152.

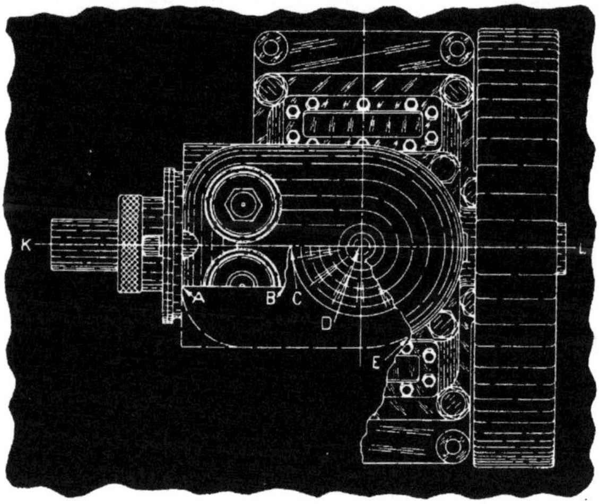

Fig. 152

The cross sections of the Cylinder as seen in the side elevation of Fig. 153 is to be known when looking at the side elevation as taken from the imaginary top plan of Fig. 152. It may be imagined to be like that shown in Fig. 154.

It is seen in the imaginary top plan of Fig. 152, that the cross section of the side elevation of Fig. 153

taken from the lines "A-B-C-D-E," which are the same lines "a-b-c-d-e" of Fig. 154.

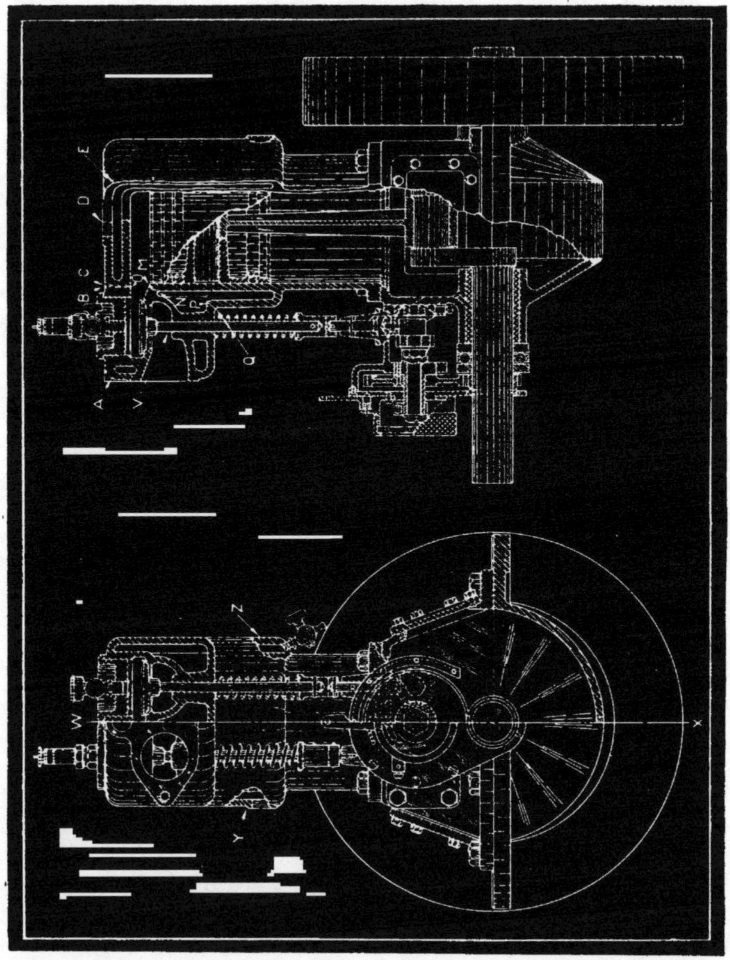

Fig. 153

The lines "A-B" and "C-D" of the top plan of Fig. 152 are parallel to each other, thus showing that the portion of the top plan of the Engine that fits to the line "A-B" of Fig. 152 is on the vertical plane, while that which fits to the line "C-D" is on another vertical plane. When these two planes are seen in the blue print drawing of Fig. 153, they are seen only on one plane, because the two separate planes are **projected** to the one plane to the side elevation of Fig. 153 as a cross section, and the only reason why it can be seen in the side elevation, is that shown of the broken lines "B-O," "M-N" and "P-Q."

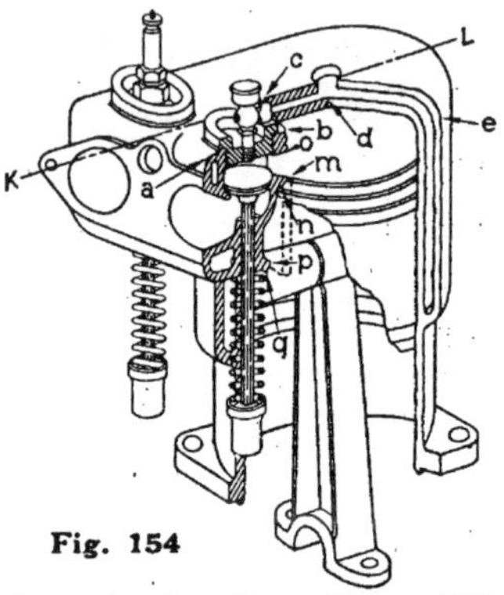

Fig. 154

When broken lines as "B-O," "M-N" and "P-Q" are seen as in the side elevation of Fig. 153, they show by the cross hatching in the cross section, a separation of two surfaces not on the same plane. The cross section as separated with broken lines "B-O," "M-N" and "P-Q," would necessarily be on two planes, as in Fig. 154.

Figure 154 also shows that the broken lines "B-O," "M-N" and "P-Q" of the side elevation of Fig. 153 are on the line "A-B" plane of the top plan of Fig. 152. That part of the Cylinder that is separated by the broken lines "B-O," "M-N" and "P-Q," is that which is on the center line "K-L" plane of Fig. 152 top plan of the Engine.

Any such broken lines as "B-O," "M-N" and "P-Q" running through any cross section of any blue print drawing similar to that shown in the side elevation of Fig. 153, shows that what is on one side of the broken line is on a different plane from that shown on the other side of the broken line.

The side elevation of the Cylinder of Fig. 153 is a difficult blue print drawing for the average mechanic to understand, as a top plan as in Fig. 152 should have been placed above the side elevation and included in Fig. 153 so as to aid the understanding. While such a top plan would aid considerably, it is not really necessary to one who is versed in the fundamental principles of reading blue print drawings, because that which is not shown, can be analyzed by giving each elevation a thorough study.

Should it be desired to analyze a cross section of the Cylinder of the side elevation of Fig. 153, would be to note that the Valve Stem "V" is shown to be on a center line of a cross section.

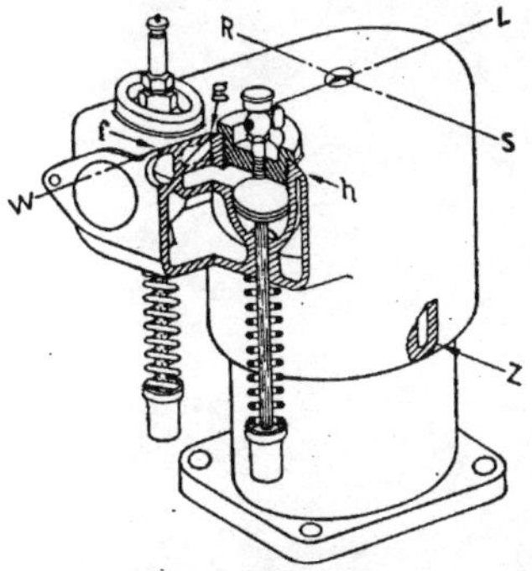

Fig. 155

There is only one Valve Stem "V" that can be seen in the side elevation of Fig. 153, which at first glance, seems to be on the Engine's center line, but is not, as you will note when you look at the end elevation of Fig. 153, for the center line "W-X" of the end elevation is seen as the natural center line of the Engine on one side of the Valve Stem "V." Therefore the Valve Stems "V" cannot be on the Engine's natural center line, but must be on another center line as shown in the top plan of Fig. 152, and seen in Fig. 154.

In the end elevation of Fig. 153, there is shown another problem of a cross section which is also to be imagined from the top plan provided as in Fig. 156.

The cross section of the top part of the Cylinder in the end elevation of Fig. 153 is taken from the center line "W-X," which is the same as the center line "W-L" shown in the top plan provided in Fig. 156.

What is shown of the cross section of the end elevation of Fig. 153 at the top part of the Cylinder should be looked for in the top plan of Fig. 156. What is shown taken from the part "F-G" of the center line "W-L" as far as "G," and then from "G" as far as "H" of the line "G-H," which is the line "f-g-h" of the perspective drawing of Fig. 155.

The broken out cross section "Z" of the end elevation of Fig. 153 must be on a center line, because all cross sections are taken from some kind of a center line. If you look for the center line "R-S" as shown on the top plan provided in Fig. 156, you can get the

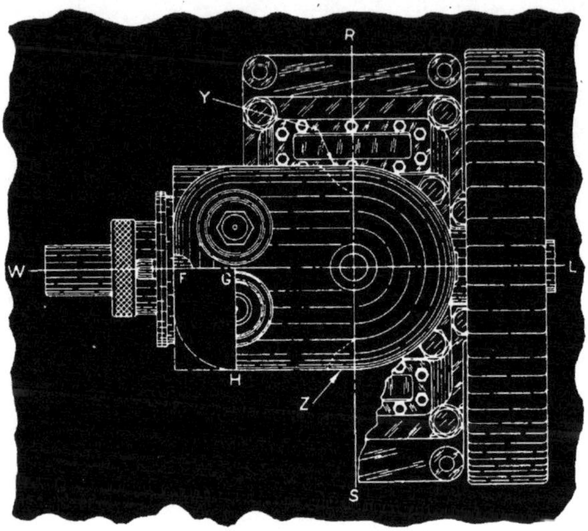

Fig. 156

location of the cross section "Z" in the perspective drawing of Fig. 155.

The cross section "Y" as shown also in the end elevation of Fig. 153 is taken from the center line "R-S" of the top plan provided in Fig. 156, but is not shown in the perspective drawing of Fig. 155, because it is on the side of Fig. 155 that cannot be seen.

The bottom part of the end elevation of Fig. 153, shows an outside half of the Engine, while the other half shows a quarter cross section taken out and cross sectioned. This cross section is not explained, but left to you to solve. The bottom part of the side elevation is also left for your study.

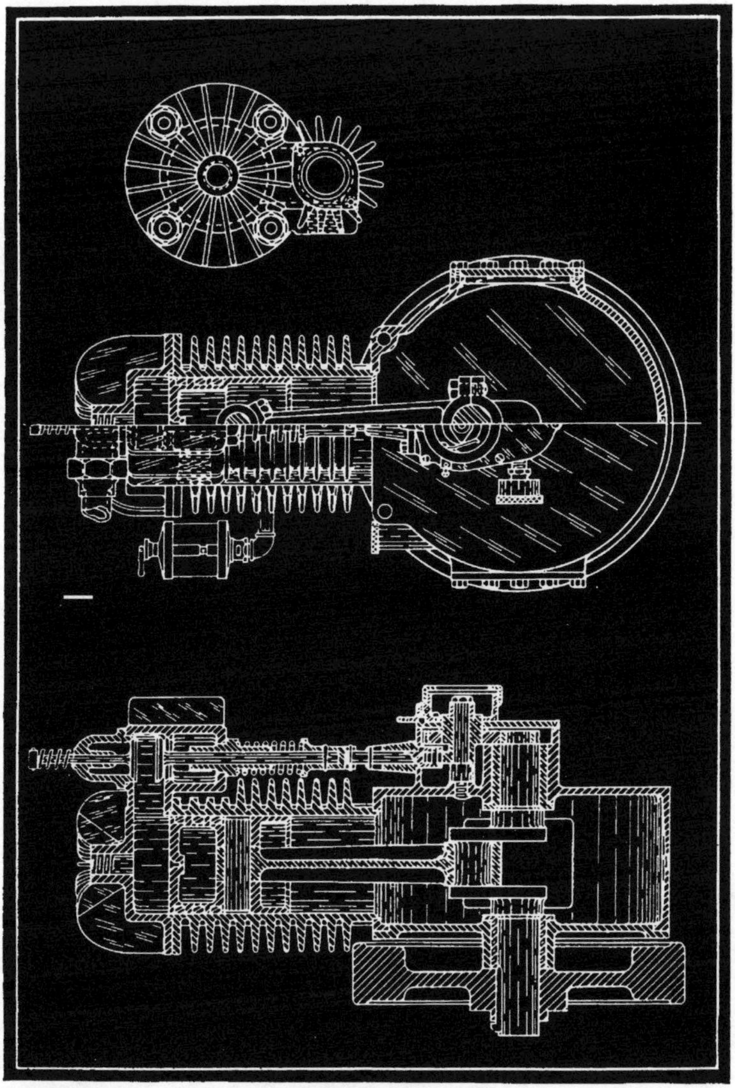

Practice Reading Figure

HALF ASSEMBLY BLUE PRINT DRAWING

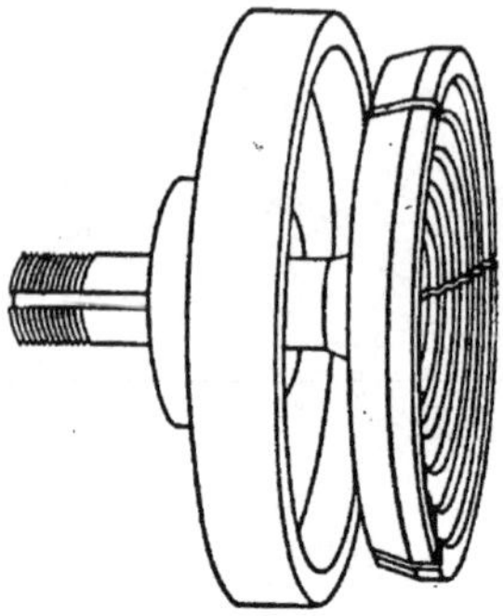

Fig. 157

There are times when a blue print drawing shows only a **half assembly,** as shown in Fig. 160.

Such **half assembly blue print drawing,** as shown in Fig. 160, is in principle like the half detail blue print drawing explained from Figs. 79 to 84.

The **half assembly blue print drawings** are generally made to save time, when the draftsman is in haste, or has only one to make up, and needs no full blue print drawing for future use.

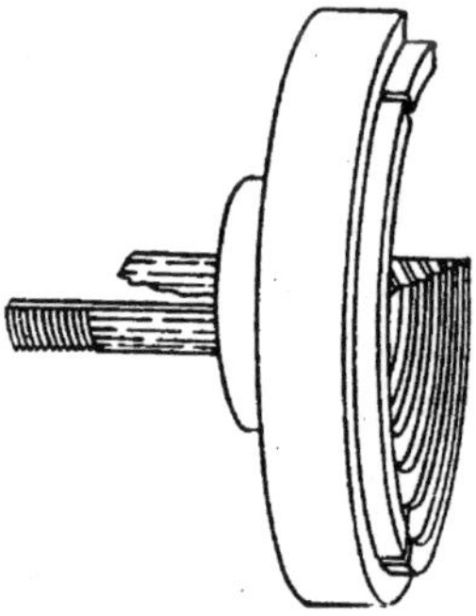

Fig. 158

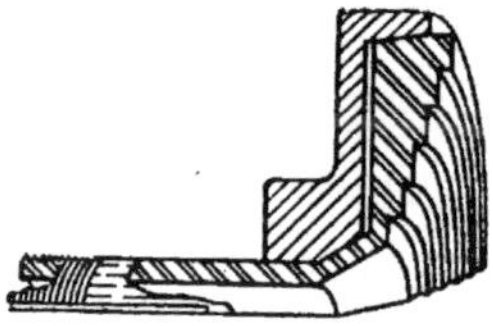

Fig. 159

Figure 160 gives all necessary information of the design.

When such words as "HALF DETAIL" or "HALF ASSEMBLY" are given below in an assembly blue print drawing, they suggest that only one half is drawn, and that the other half is to be imagined just as that drawn in Fig. 160, and realized with the dashed lines that the half assembly of Fig. 161 shows.

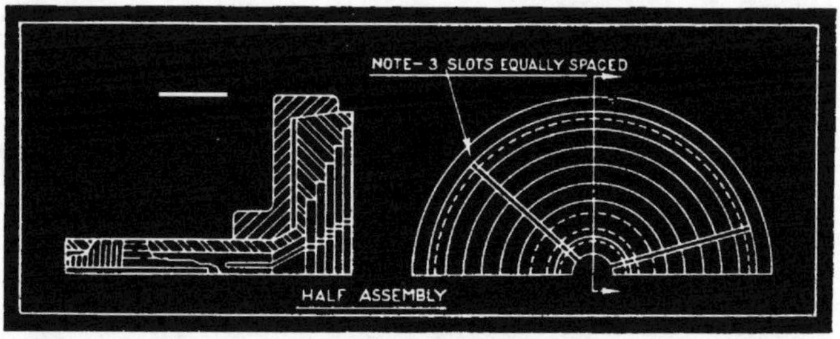

Fig. 160

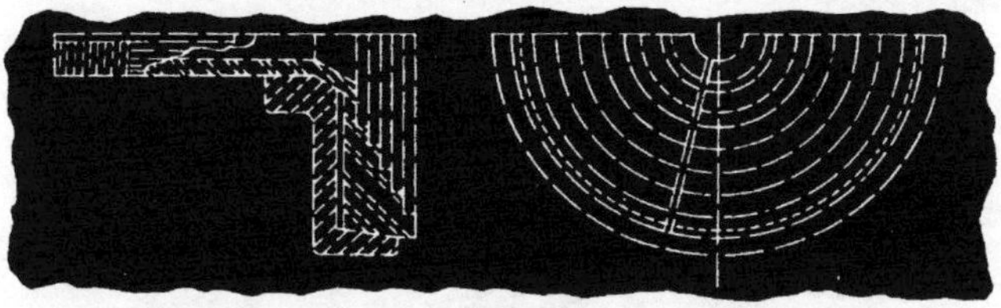

Fig. 161

What is seen in the cross sectioned **half assembly blue print drawing** of Fig. 160 can be understood from Figs. 157, 158 and 159.

Figure 159 represents what is seen of the L-H. end elevation of Fig. 160 as on the center line of Fig. 158.

PART VIII

GROUP DETAIL BLUE PRINT DRAWINGS

In Fig. 166, there are several separate detail parts all drawn in one **group blue print drawing.** Each detail in this group must be studied apart from the others shown, so that each will be understood.

The object of giving so many details in one **group** is to remove the necessity of making a separate blue print drawing detail of each part. Another reason is

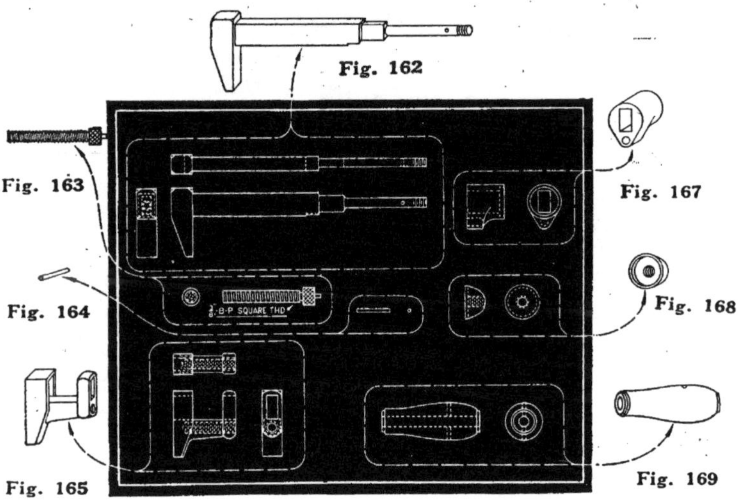

Fig. 162

Fig. 163

Fig. 164

Fig. 165

Fig. 167

Fig. 168

Fig. 169

Fig. 166

that all the parts may be before the mechanic so that the mechanic does not have to scout through several blue print drawing details before he can find the detail that he wants.

A disadvantage of a **group detail blue print drawing** as shown in Fig. 166 is that, should there be only one blue print drawing furnished, only one mechanic could conveniently use it at a time. This would not allow the free use of same without interference of the other mechanics, should they be working on the other part of the group at the same time.

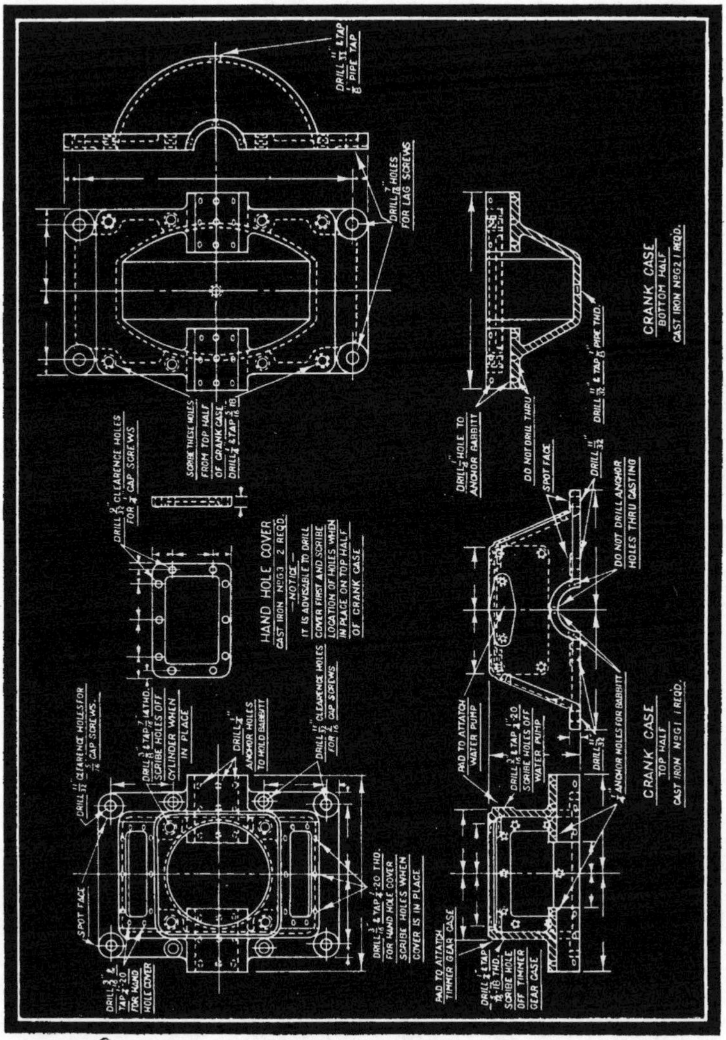

Fig. 170

The **group of details** as shown drawn in Fig. 166 for each separate part of the Monkey Wrench, as that encased with a dashed arrow line, may be understood from the perspective drawings of Figs. 162, 163, 164, 165, 167, 168 and 169.

In Fig. 170 is a **group of details** containing remarks about the parts; as is true of the general run of all blue print drawings. The group of details of Fig. 170, is for study, without any perspective drawing as was used in the other explanations of this book, to show what each part is like.

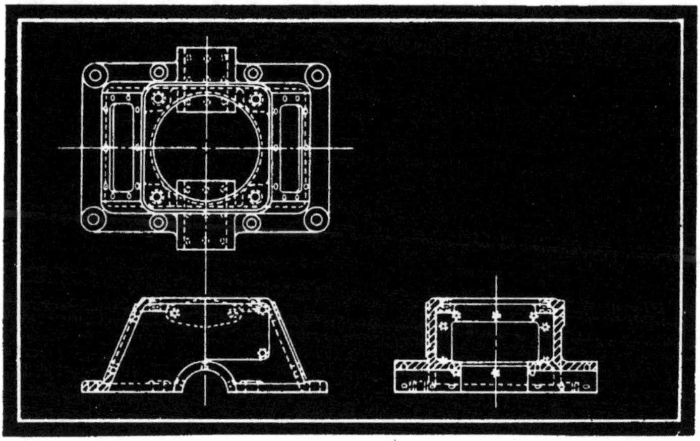

Fig. 171

The **group of details** of Fig. 170, is left to you for study and practice reading.

Figure 170 is a typical blue print drawing, giving place between the dimension arrows for the sizes, and the general remarks that are always shown of a detail.

Another **group detail blue print drawing** is shown in Fig. 130, as well as on **page** 119.

SEPARATE DETAIL BLUE PRINT DRAWINGS

The blue print drawings of Figs. 171 and 172 are used to show separately, the details as composed in either of the blue print drawings, and likewise have no perspective drawing to help the reader to an understanding, but left for you to study for their meaning.

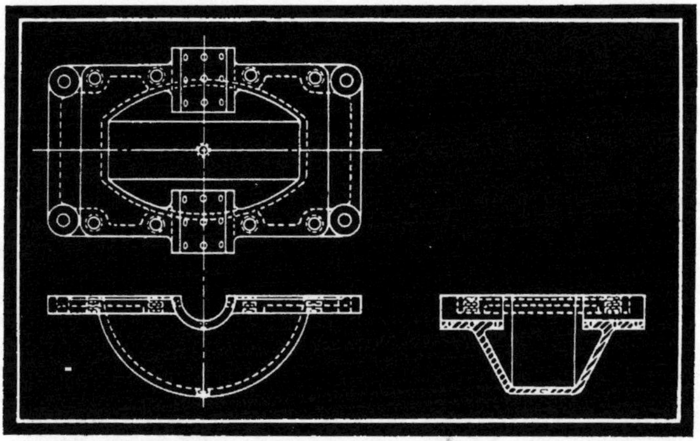

Fig. 172

The **separate blue print drawing details** as shown in Figs. 171 and 172 may be composed from one to many more elevations and plans, all depending upon what the blue print drawing is to show.

It may be to your interest to know that Figs. 170, 171 and 172 were taken from the assembly blue print drawing of the Engine of Fig. 153.

DIAGRAM BLUE PRINT DRAWINGS

In Fig. 174 is shown a **diagram blue print drawing** for the wiring of a Gasoline Engine provided with electric wires for the electric spark.

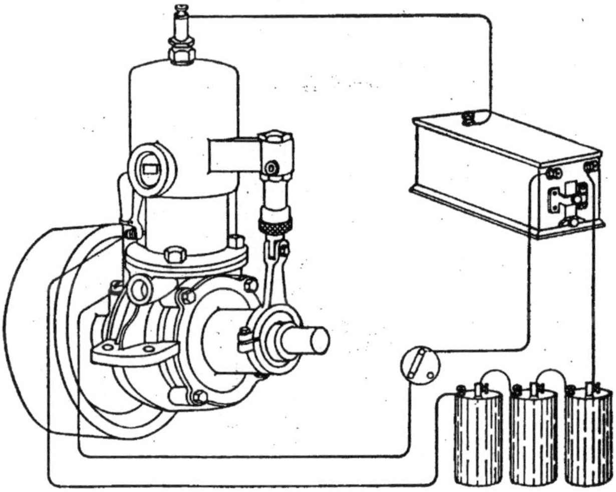

Fig. 173

The **diagram blue print drawing** of the Engine of Fig. 174, is the proper way of showing how the wires are to be connected.

You will note that the center Lever "S" of the Engine of Fig. 174, is drawn with full lines, and that the two Levers as shown with the letter "T" are drawn with dashed lines. These two Levers "T" drawn with dashed lines are to show the limit of the swinging positions that the full drawn line Lever "S" can be placed in, **and not,** as they might seem to show, that they are three Levers.

Figure 173 gives a good understanding of the wiring as shown in the blue print drawing of Fig. 174.

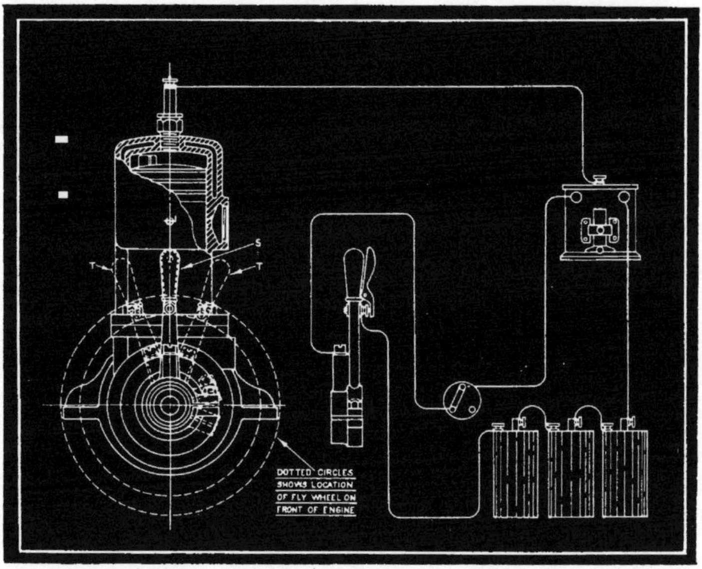

Fig. 174

GEAR BLUE PRINT DRAWINGS

Blue print drawings of Gearing are quite simple to understand and, providing that you understand the principles of **projection** of elevations and plans as treated in the first part of this book.

The object of showing blue print drawings of Gearing is to show the principles of teeth as drawn on same.

It is not necessary to explain in this book any remark given on a blue print drawing of any kind of a Gear which relates to the mechanical design and usage. Such remarks are understood as part of the trade of the mechanic who has anything to do with Gears.

SPUR GEAR BLUE PRING DRAWING.

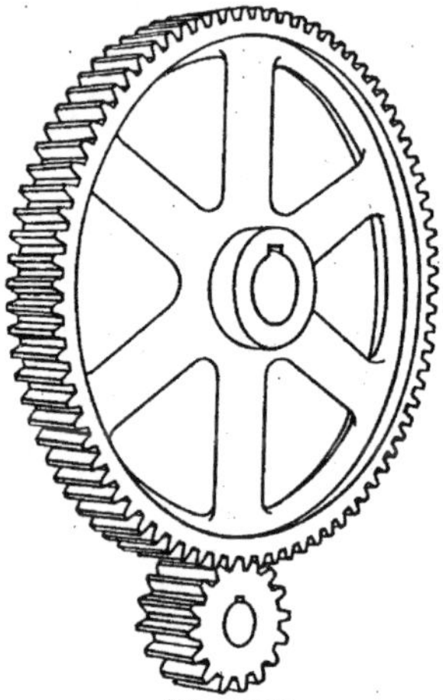

Fig. 175

Figure 175 shows a perspective drawing of two Spur Gears in mesh for an easier understanding of the blue print drawing of Fig. 176.

A blue print drawing of a Gear usually shows a few teeth drawn in the outer rim, as marked with the letter "T," as in Fig. 176.

It was not necessary for a draftsman to draw all the teeth around the rim on the side elevation of Fig. 176, because showing all the teeth would only make drawing nice to look at. Nothing more would be shown of the Gear than is shown by the few teeth with the letter "T."

Even the few teeth shown with the letter "T" in the side elevation of Fig. 176 are not necessary on a blue print drawing of a Gear, because with the pitch circle drawn, besides other marks or words, such as "P.D." after the pitch diameter, the diametral pitch, etc., that are usually on a blue print drawing, there is enough information to know that the drawing is a Gear.

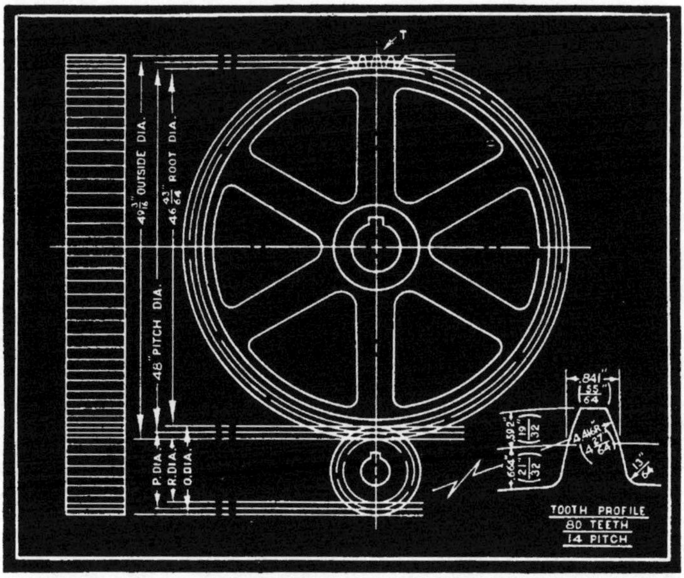

Fig. 176

Another reason why no teeth need be drawn to show that a blue print drawing is a Gear, is that there would be such information in the form of reading matter on a blue print drawing to show that it is a Gear, such information as "80 TEETH"—"14 PITCH," and possibly the circular pitch, given in Fig. 176 along with other words.

The tooth profile as shown in Fig. 176 blue print drawing of the Gear, need not necessarily be given, but such information always adds considerably to the understanding whenever shown.

The end elevation of Fig. 176 need only show with straight lines where the teeth are to be. No actual teeth need be shown, as is shown with the letter "T" in the side elevation.

SPIRAL GEAR BLUE PRINT DRAWING.

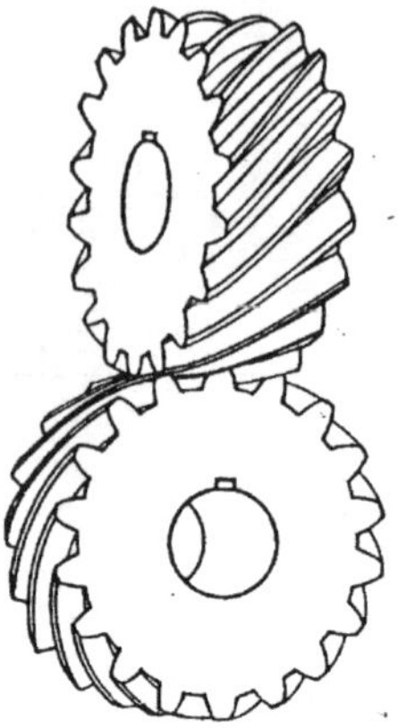

Fig. 177

Figure 178 is a blue print drawing of Fig. 177 that shows a set of Spiral Gears.

What has been explained of Spur Gears may be said of Spiral Gears.

The slant lines "S" of the **Spiral Gear blue print drawing** of Fig. 178 suggests the location of the teeth, like in the **Spur Gear blue print drawing** of Fig. 176.

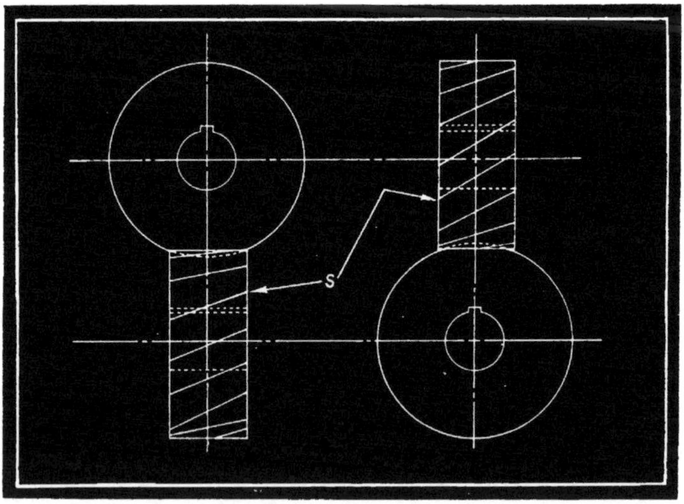

Fig. 178

BEVEL GEAR BLUE PRINT DRAWINGS.

Figure 179 shows a perspective drawing of two Bevel Gears in mesh, which when drawn as a part of

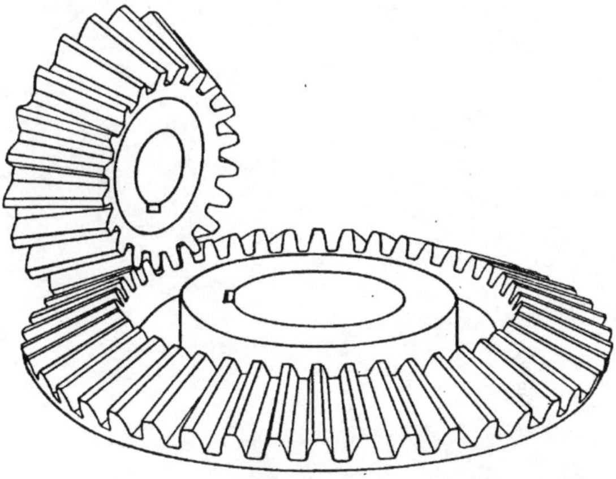

Fig. 179

Fig. 180

an assembly blue print drawing, could be like that of Fig. 180.

In the blue print drawing of Fig. 180, there is but little information given of the shape of the Gear drawn. What is given is sufficient for the purpose of showing the Gear located in a portion of an assembly blue print drawing.

Fig. 181

Quite often a pair of Bevel Gears are only partly cross sectioned, and partly drawn to represent as they actually look, like the blue print drawing of Fig. 182 representing the Gears as shown in Fig. 181.

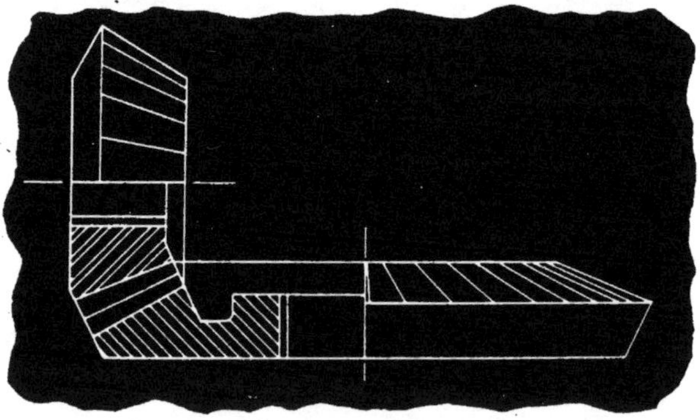

Fig. 182

A blue print drawing, such as that in Fig. 182, providing no dimensions are needed, offers two ways for the mechanic using blue print drawings to understand it.

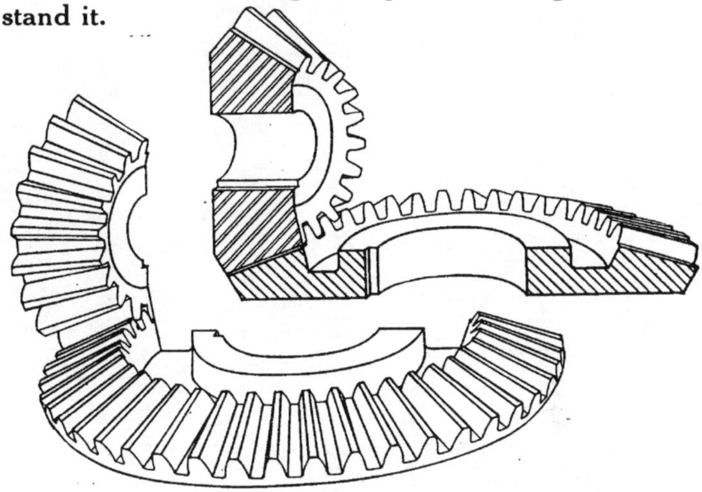

Fig. 183

For a detail of the pair of Bevel Gears shown in Fig. 183, a blue print drawing like that in Fig. 184 is the most common and practical way. In Fig. 184, details of the two Bevel Gears are shown by the use of a cross section in the blue print drawing, to which dimensions can be easily applied.

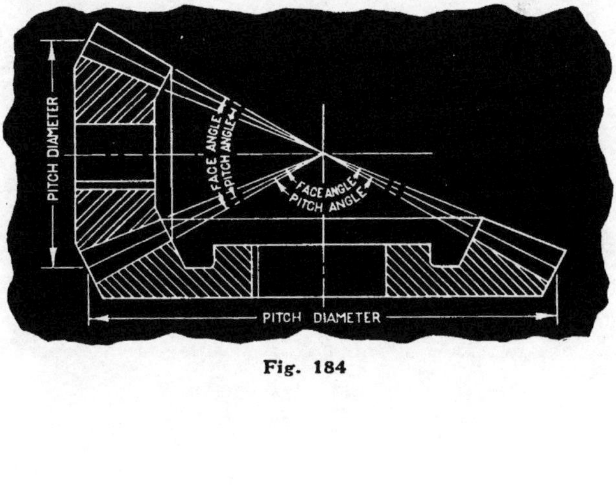

Fig. 184

WORM GEAR BLUE PRINT DRAWING.

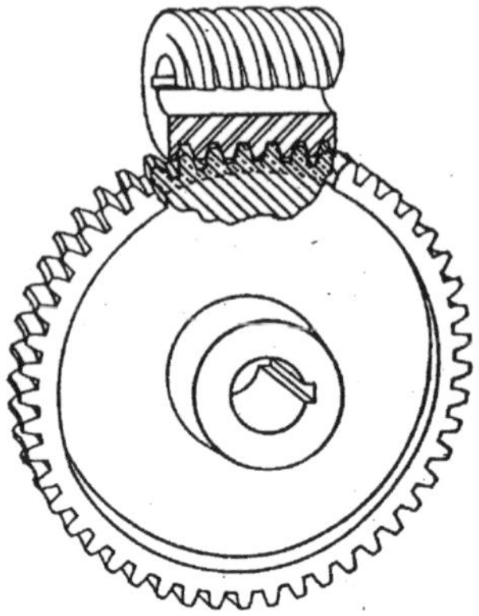

Fig. 185

Figure 186 is a blue print drawing of the Worm and the Worm Gear shown in the perspective drawing of Fig. 185.

The end elevation of Fig. 186 shows in the bottom half, the slanted position of the teeth of the Worm Gear, besides the quarter cross section of the upper half, while the side elevation of Fig. 186, shows a broken out cross section of a part of the Worm and the Worm Gear, easily understood by looking at Fig. 185.

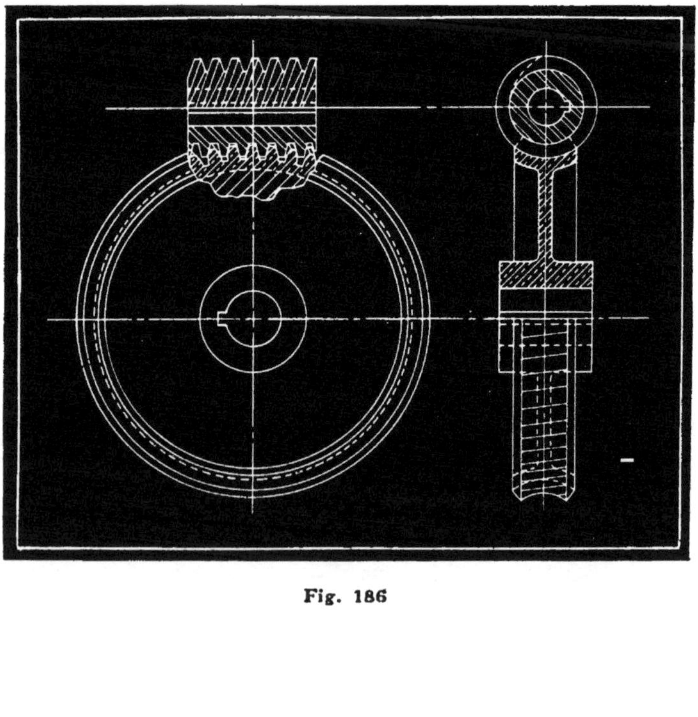

Fig. 186

MOVEMENT OF TRAVEL IN ASSEMBLY BLUE PRINT DRAWINGS.

In Fig. 187 is shown the **movement of travel** of a Governor so as to represent on a blue print drawing the path of travel and the limit of space such movement occupies.

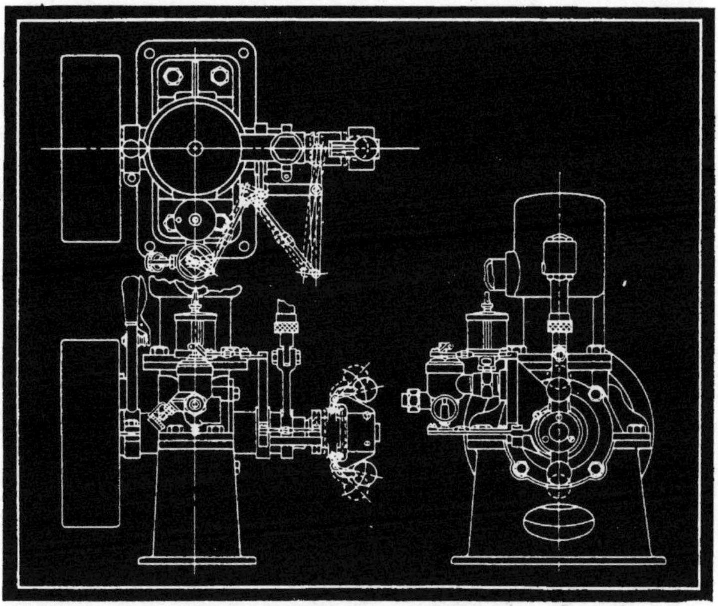

Fig. 187

Such **movement** of **travel** is practically only shown in an assembly blue print drawing with dotted lines, as is shown in Figs. 174 and 187.

Sometimes the **movement of travel** is shown with the center lines only, **and not** with dotted lines outlining the shape of the part traveled.

The distance of any **movement of travel** is measured between center lines on the center of the arc or path of travel.

PART IX

CONVENTIONS

Conventions when applied to the reading of a blue print drawing is the necessary words or symbols to further a complete understanding of the blue print drawing.

Practically every blue print drawing contains **conventions** in some form, which are to be understood as a part of the actual drawing.

A **convention** may be more fully understood as a particular mark, word, design, symbol or sign which has a general meaning as adopted in the science of blue print drawing, which when seen, is to be understood for its full meaning.

The following symbols, letters or words when appearing on a blue print drawing are called **conventions.**

MATERIAL MARKS

This chapter explains the **conventions** of materials shown with **initials** on a blue print drawing, are also the symbols of materials as shown in Fig. 89. If no **initials** are used to show what the material is, then the full name is shown on the blue print drawing. Such **initials** and their meanings are as follows:

C.I. means	**Cast iron.**
C.S. means	**Cast steel.**
C.R.S. means	**Cold rolled steel.**
Cp. means	**Copper.**
W.I. means	**Wrought iron.**
Mal. means	**Malleable iron.**
Br. means	**Brass.**
T.S. means	**Tool steel.**
H.S.S. means	**High Speed** steel.
Alum. means	**Aluminum.**

SCALES

In the making of anything from a blue print drawing, it is understood that the blue print drawing as a general rule is drawn to a **scale** or stated size. Should the article to be drawn be not too large, and if it can be drawn on a moderate size sheet, a **full size scale** is almost always used, and such is to be stated in the blue print drawing as "SCALE—FULL SIZE" or "SCALE —12″ = 1′." Whatever the desired **scale** may be, it is to be stated on the drawing.

Should a very large article, such as a Steam Roller or any of its large parts have a drawing, such a drawing could not be conveniently made to **full size,** therefore a **scale** of considerable reduction would be shown, such as "SCALE—12″ = 1″," etc., meaning that every inch in length, width or height drawn with such a **scale** would represent one foot.

There are many **scales** that can be used in the making of a blue print drawing, all depending upon the relation of the size of the part to be made, to the size of the drawing sheet used.

The following **scales** are most commonly used, although there are **scales** other than those listed here that are used:

6″	=1′ means	1/2	size	¾″=1′	means	1/16	size
4″	=1′ means	1/3	size	½″=1′	means	1/24	size
3″	=1′ means	1/4	size	⅜″=1′	means	1/32	size
2″	=1′ means	1/6	size	¼″=1′	means	1/48	size
1½″	=1′ means	1/8	size	⅛″=1′	means	1/96	size
1″	=1′ means	1/12	size	1/16″=1′	means	1/192	size

When an article to be drawn is very small in actual size, an **enlarged scale** is generally used so as to show everything much larger than it actually is, so that ease and accuracy can be used by the draftsman in the making of such drawing, therefore a **scale** such as "DOUBLE SIZE," "SCALE—6″—1″," etc., would be used in the making of such drawing.

Practically every blue print drawing is made to a **scale,** and whatever **scale** is used is most generally stated on some part of the drawing.

Sometimes a **scale** is not noted on a blue print drawing, so as not to permit any freedom in taking dimensions from the drawing. Should a dimension be omitted or will not check up, it is always advisable to take up such matters with those who are responsible for such drawings.

In almost any blue print drawing, even when the design is practically finished, a minor change may be necessary. Such a change may be taken care of mostly in dimensions. The mechanic should not **scale** such a drawing. For this reason, almost all blue print drawings show the words "DO NOT SCALE."

Whenever the words "DO NOT SCALE" are shown on a blue print drawing, never measure any dimension from the drawing, but go only by whatever dimensions are shown.

Another reason why the "DO NOT SCALE" is so often shown on a blue print drawing is because of the inaccuracy of the drawing due to shrinkages and stretching of the drawing when several days have been required to make it. Such conditions are taken care of, and allowances made in the draftsman's calculations when he states the dimensions on the blue print drawing.

MARKS

f means This mark on any surface drawn is to show that the surface is to be machined, in other words, **to finish** smooth, as Fig. 188 shows.

ff means To be finished **extra smooth.** When the initials **F.A.O.** is used, no **f** need be shown.

F.A.O means To **finish all** over.

G. means To **grind.**

GEARING MARKS

P.D. means **Pitch diameter.**
D.P. means **Diametrical pitch.**
O.D. means **Outside diameter.**
R.D. means **Root diameter.**
C.P. means **Circular pitch.**
C. means **Clearance.**
C. to C. means **Center to center.**

PATTERN NUMBERS

PATT. means **Pattern number.**

When a number is shown such as PATT. K-12 or PATT. 1933-C-22, on a blue print drawing, it is to be understood that a particular method is used in marking the pattern for record keeping.

PATT. K-12 is a number applied to a pattern, in which the letter "K" shows the series, and the figure "12," is the order in which it belongs.

PATT. 1933-K-22 would show that the figure "1933" would be the blue print drawing number, the letter "C," the size of the drawing, and the figure "22" would be the twenty-second detail of the drawing "1933." The pattern number would be the same as the drawing number for one record to keep tab of.

CROSS HATCHINGS

Any **cross hatching** of any blue print drawing is a convention, because it is used to give a general understanding among the readers of blue print drawings.

In Fig. 188 is shown a five spoked Hand Wheel, with an R-H. and a L-H. end elevation to its side elevation.

As previously stated in the first part of this book, only one of the end elevations was needed as a **projection** from the side elevation; but in Fig. 188, the using of both, the R-H. and the L-H. end elevations in the same drawing is only to show by contrast the effect of a **practical projection** in relation to its **theoretical projection.**

The R-H. end elevation of Fig. 188 is the **theoretical projection** from the side elevation of what is seen of the spoke below the center line "A-B" when such spoke does not completely appear on the center line "C-D."

Leading authorities claim and always show on their drawings, that when a spoke of a Wheel is not com-

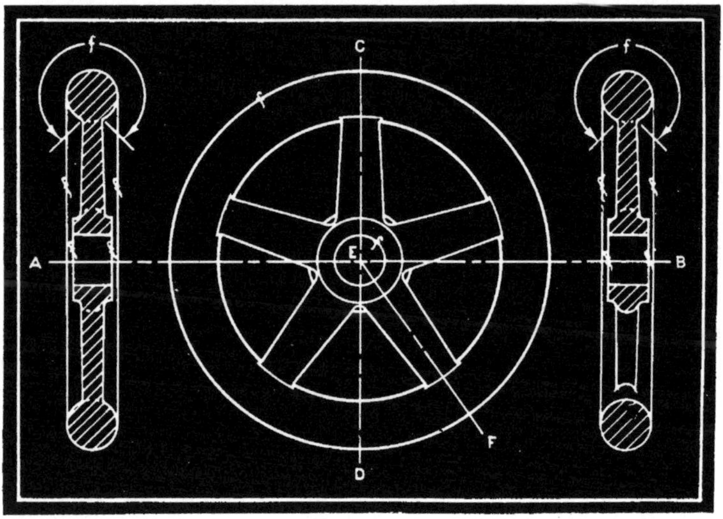

Fig. 188

pletely on the vertical center line of the side elevation, that such spoke is to be **projected** as a center line **projection,** as the L-H. end elevation shows, **and not** as the R-H. end elevation shows.

The writer cannot understand, nor has ever been able to find out the "WHY" of this so-called **practical spoke projection,** which applies only to the **projection of spokes,** but knowing its general use, imparts to you the condition as it exists.

Such a blue print drawing of Fig. 188 requires only one end elevation, and such end elevation would no doubt be as the L-H. end elevation shows, although there are many end elevations of the spoke as the R-H. end elevation shows.

Ordinarily, such a cross section as shown in the L-H. end elevation of Fig. 188 would be considered a zig zag cross section taken from the line "C-E-F" of the side elevation, which would also have the words stated below the L-H. end elevation stating such.

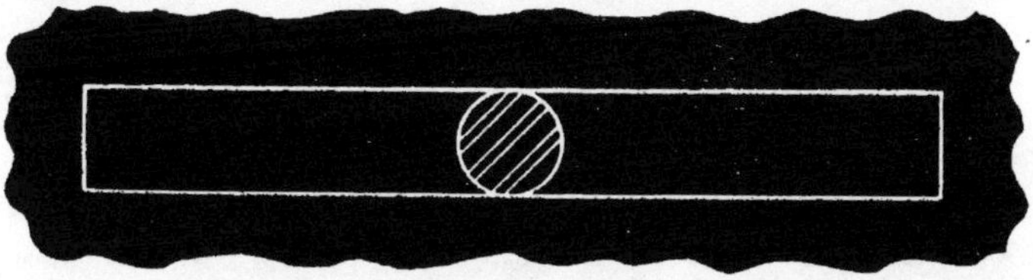

Fig. 189

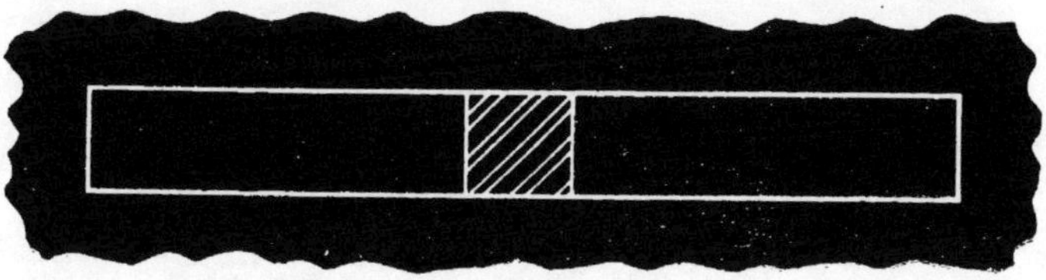

Fig. 190

Many articles are drawn with only one elevation. To show its other elevation, requires only a cross section of whatever shape it may be, cross hatched somewhere inside of the same, as is shown with Figs. 189 and 190.

LINES

Lines of blue print drawings are made with various meanings, which are as follows:

Full lines

For all outlines of an article as drawn.

Invisible, dotted or hidden line.

The invisible or hidden line is commonly called the dotted line which is used in a blue print drawing wherever there is a **line hidden.** Such an invisible or hidden line is also used to show the movement of travel of parts, as is shown in Figs. 174 and 187.

Center line

Dimension or extension line.

The use of dimension lines as shown, is rarely ever used, for the regular medium line with arrows placed on each end is what is almost always used.

The above line is practically used only as an extension line.

CENTER LINES

Practically every blue print drawing has one or more **center lines,** which are used to show the centers of an article.

Should an article drawn have two or more **center lines,** then such **center lines** show each center construction of the most particular part to work to, as almost all of the important dimensions are taken from a **center to center** measurement.

Holes are always measured between **center lines** which show the **center to center** dimension.

Whenever such letters as **"C to C."** are shown on a blue print drawing, they are to be understood as the **center to center** of center lines.

Such symbol as this appearing on a **line** always means **center line.**

DIRECTIONS

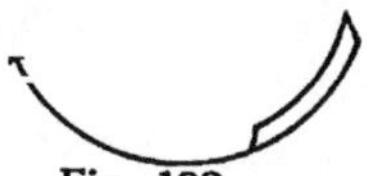

Fig. 191 Fig. 192 Fig. 193

Straight direction arrow = as Fig. 191 shows.
Centrifugal direction arrow = as Fig. 192 shows.
Both directions arrows = as Fig. 193 shows.

DIMENSIONS

Dia. means **diameter.**

The letter "**R**" of Fig. 194 means **radius** when shown after a **dimension,** and is inside of an arrow line. It shows that the **dimension** is measured from the center of a circle or arc to its outside edge, on the blue print drawing, as in Fig. 197.

Fig. 194

Fig. 195

Figure 195 shows a **radius dimension,** explanation of which is the same as the above. only that it is folded. The idea of such a **folded radius line** is to show that the actual center is in a straight line from the straight part of the arrow line that the **dimension** is to be measured from. The reason why the actual **center** is not shown in its place, is to make room for other details that can be drawn in its place. By **folding** the radius arrow line, such condition can be provided for. Such a radius arrow line is shown in Fig. 176.

$$12.875''\frac{+.005}{-.003}$$

When such **fractional dimensions** as given above are shown in decimals after a whole number and decimal, it is understood that the allowable variations of the **dimension** is not to be over five thousandths inch oversize, or three thousandths inch undersize. The arrangement of the above **dimension** can be in any combination other than that shown.

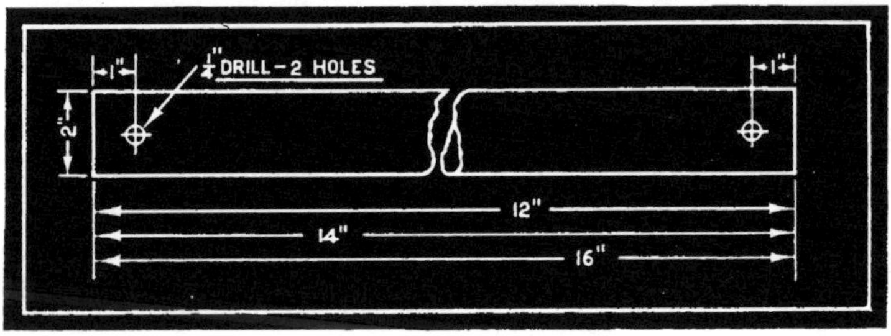

Fig. 196

When two or more **dimensions** are seen between the same limit, they are to show that there are as many sizes to be made as there are **dimensions,** as in Fig. 196.

DIMENSION CHANGES

Practically every blue print drawing, after being in use a little while requires **dimension changes** to meet new conditions.

When a **dimension change** is required, it is generally the practice of draftsmen to erase the dimension figures and insert in their place a capital letter inclosed in a circle, and whatever letter is used refers to the change column of the drawing, showing what the dimension was, what it is now to be, the date that such a change was made, the authority of the change, and any other necessary remarks. Such change column would be somewhat like Fig. 197.

Any change in the design, other than dimensions or remarks would necessarily require a new drawing, and such would not come in the scope of the reader of the blue print drawing.

TITLE BLOCKS

Practically every blue print drawing has a **title block** somewhere in the drawing, and most generally on the lower right hand corner, as in Fig. 197.

Inside of the **title block** is all the information that is required of those responsible for the drawing. This information is rarely of much use to the general reader of the blue print drawing, except for the **name** of the **article** drawn, the **scale** used, and possibly the **date** when the drawing was made.

Scales are seldom ever placed inside of the **title block** of a group detail, for the reason that the drawing may have several different scales to show for each detail, which would make it impossible to state conveniently the scale in the small space allowed, therefore each detail in a group of details generally shows the scale placed somewhere below the detail as drawn.

In every separate detail blue print drawing, the scale is generally shown in the **title block** as in Fig. 197.

All of the blue print drawing in this book, except Fig. 197, have no **title block,** for it was not necessary to consider such in the educational problems dwelt with.

BILL OF MATERIAL

In every blue print drawing, there is some sort of material such as Screws, Taper Pins, Babbit, etc., needed for the completion of the article drawn, and such material is usually, but not always, listed in the upper right hand corner of the blue print drawing. Such list of material is called a **bill of material,** as in Fig. 197.

In assembling all of the details of an article, there is material such as Grease Cups, Gauges, etc., required for use then, and not when the detail is being made up. In order that all of the miscellaneous materials may be ordered and kept in readiness until needed, a separate **assembly bill of material** is almost always made up for such conditions.

In Fig. 197, the **bill of material** gives in the column, only one article that is needed for the detail to be made. Should any other material be needed, a place is provided for the same.

SPECIAL SECTION DETAIL

When any blue print drawing is made small, and a particular part of such a drawing has enough importance attached to it, such particular part may be extended out from the detail of the part, as in Fig. 197.

There are two **special section details** in Fig. 197, the **upper one** is to show the dimension of the slot without showing a whole top plan, while the **lower one** is an extended out cross section drawn double size to show the ribs, and where the web is recessed for a clearer view.

POSITIONS

In detail blue print drawings, it is advisable for draftsmen to show **positions** in dashed lines, such as the end of the Rod, the Key and the other half of the Eccentric Strap, as in Fig. 197.

The showing of **positions** in Fig. 197 offers the reader of the blue print drawing an understanding of the actual detail that is to be made up by being drawn in full lines, while the relation of the detail is shown for its **position** with the other details that it is placed with.

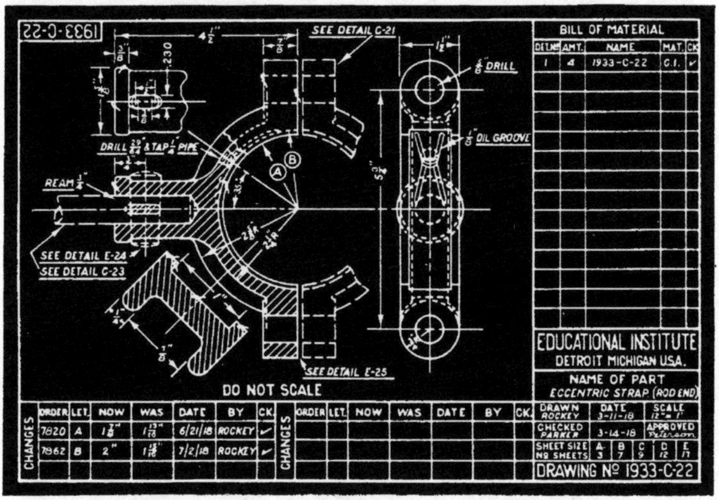

Fig. 197

Other blue print drawings are made for the detail of the **positions,** which in Fig. 197 is shown with the words "SEE DETAIL C-21," —"C-23," —"E-24" and "E-25."

The Key as shown cross hatched in Fig. 197 is in its natural **position** according to its end elevation **projection,** but when shown as it is with dashed lines in an upright position at 90° to its actual **projection,** a clearer understanding of the whole Key is gained for its **position.**

BREAKS

Breaks in a blue print drawing of Shafts and Rods are always used to show what is left out, so as to shorten the actual drawing of same.

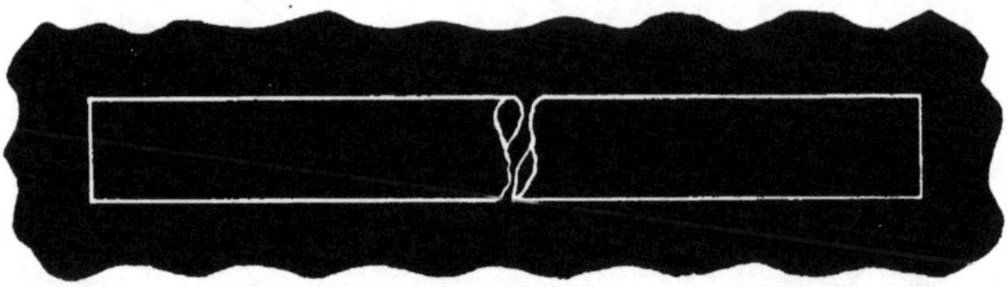

Fig. 198

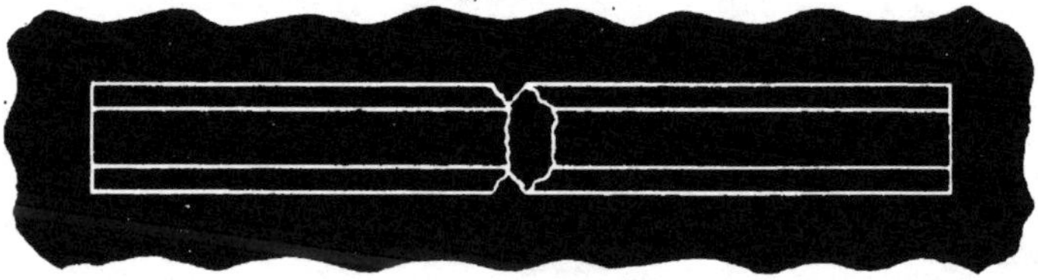

Fig. 199

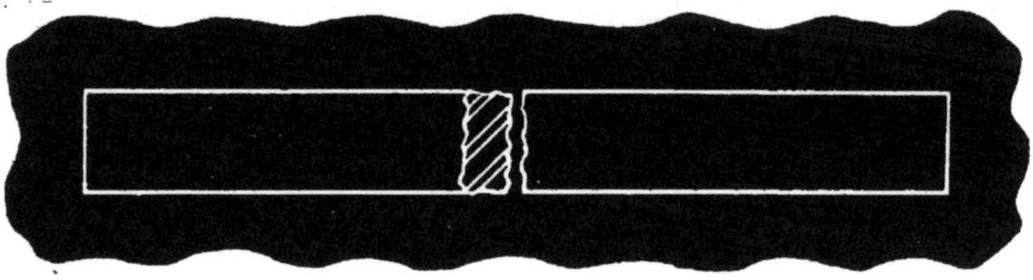

Fig. 200

Whenever a Shaft or Rod is shown broken, the **break** is made with irregular lines to give an idea of the shape, as in Fig. 198, 199 and 200. To show a **break** in other articles drawn, the same principle is used.

NUT PROJECTIONS

When Nuts are shown in an elevation, they appear according to the position they happen to occupy in the drawing as in Figs. 206 to 210.

The dotted lines showing the hole, and the dotted lines showing the thread, are seldom ever shown for a Nut of Fig. 206, when such a Nut is drawn with another article, because the shape as drawn shows that it is a Nut. Figure 207 might be drawn to show the hole, because there is little to indicate that it is a Nut.

Fig. 201

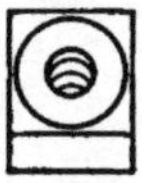
Fig. 202

Fig. 203

Fig. 204

Fig. 205

Fig. 206

Fig. 207

Fig. 208

Fig. 209

Fig. 210

It is not likely that Fig. 207 would ever be drawn to show a square Nut in a blue print drawing, for Fig. 206 may better indicate a square Nut.

Figure 206 shows a square Nut, while Fig. 208 shows a hex Nut which look considerably alike, but you will note in Fig. 206 that the square Nut is much wider than the hex Nut in Fig. 208. Further, the square Nut is considerably thinner than the hex Nut.

A **castelated** Nut of Fig. 205 and 210, is usually hexagonal, in order that its six sides may allow more slots for the alignment of the Cotter Pin, thru the Cotter Pin hole of the thread.

No standard Nut should ever be cross sectioned; it should always be drawn as a whole Nut.

SCREWS, HEADS AND THREADS

The **heads of Screws** are to be understood in the same manner as explained of Nuts, only that the width of **Screw Heads** is not as great as that of their Nuts.

Standard Screws and **Heads,** like Nuts are always drawn as a whole and never cross sectioned.

Screw threads, whether they are on a Screw, Stud, Rod or Bolt, are all drawn to convey the same meaning.

Any conventional method of showing **threads** can be used wherever a **thread** is to be drawn, as all **threads** as shown convey the same meaning. Con-

Fig. 211

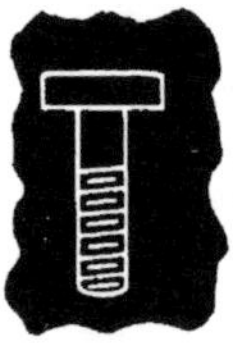

Fig. 212

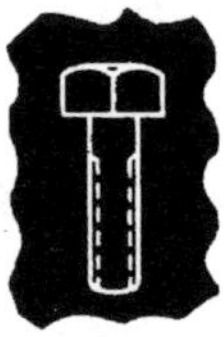

Fig. 213

Fig. 214

Fig. 215

ventional methods of showing **threads** are shown in Figs. 211, 212, and 213.

The method shown in Fig. 213 is the one used mostly by tool designers, but is rarely ever seen in a machine drawing.

All **threads** in Figs. 211, 212, and 213 are understood as being a Single United States Standard **Thread,** and should a double, triple or quadruple or as wanted, such **threads** would be drawn with a larger angle as shown in Fig. 214, and with a note to state such a condition.

When U.S.F. is shown for a **thread** it means **United States** Standard Form Thread. All other single **threads** are understood to be the "V" shape.

Fig. 215 shows the angle direction of the **thread** to be in the opposite direction for a left hand **thread.**

TAPPED HOLES

The full drawn circle with a dotted outer circle as in the top plan of the hole going thru the special Nut of Figs. 216 and 217 shows the proper way of indicating a **tapped hole.**

Some leading authorities show in their drawings, a full line circle, with an outer full line circle drawn to about 45°, or a little past the quadrant of the two center lines to represent a **tapped hole** as in the top plan of Fig. 218. This should be borne in mind.

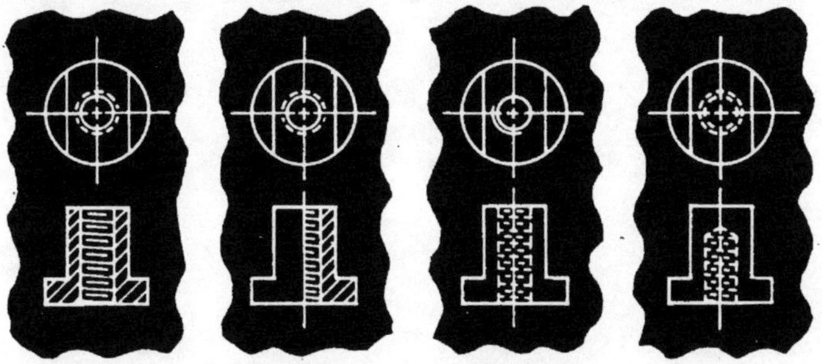

Fig. 216 Fig. 217 Fig. 218 Fig. 219

Should the **tapped hole** of the Nut not go thru, as in Fig. 219, then the inside, as well as the outside circles are both drawn dotted in the top plan to show a **hidden tapped hole.**

Either one of the elevations of Figs. 216, 217 and 218, is given as a correct way of showing a **tapped hole.** Figure 216 is the **full** cross section, Fig. 217 a **quarter cross section,** while the elevations of Figs. 218 and 219 show the **hidden method.**

Should the **threads** in question be left handed instead of right handed as they are now shown, then the angles of both, the **Screw thread** and the **hole thread** are shown reversed.

When a **Screw thread** is shown not to occupy the whole depth of the **tapped hole** as in Fig. 220, the

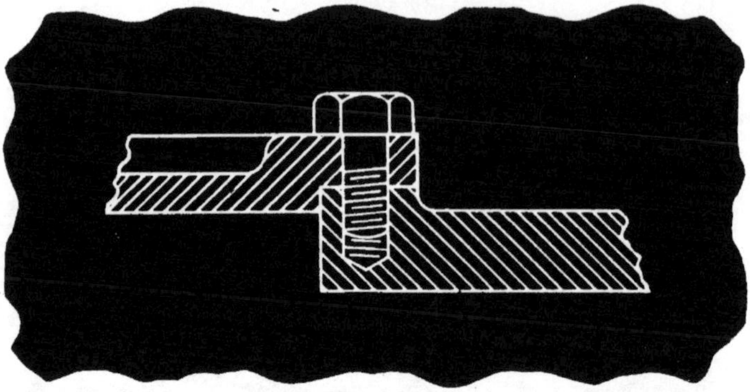

Fig. 220

angle lines showing the **thread** of the Screw are drawn as it should be shown, and the lines to show the **thread** of the **hole** are drawn to the opposite angle.

CORED HOLES

Whenever a remark as "**core**" is given on a biue print drawing, such remark means that the article to be made is first cast in a foundry and left as it is, or to be finished in a machine shop. Whatever the shape of the **core,** the casting will show a hole of that shape, to be either left as it is mounded, or other remarks will show where it is to be finished.

Should the **cored hole** be required to be finished, such **cored hole** is used to relieve excessive metal in machining.

The **cored holes** are noted in the blue print drawing of Fig. 221.

GROUPED HOLES

When there are several **holes grouped** closely together, or in uniform location about the article, it is not necessary to state the sizes and given remarks for each **hole,** for a blanket note may be used, especially

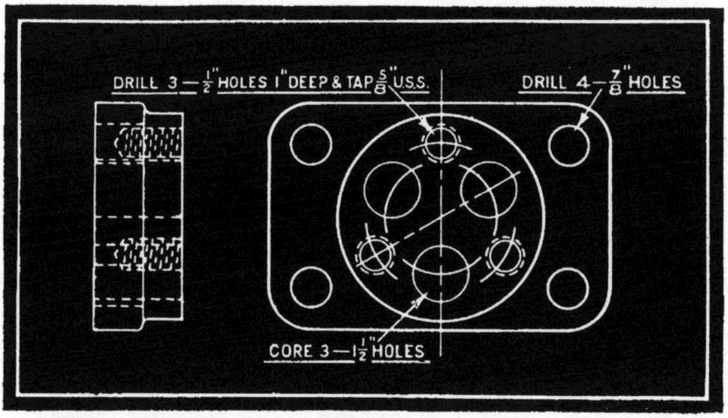

Fig. 221

if each class of **hole** is noticeably of the same size, as in the blue print drawing of Fig. 221.

BEARINGS

Figure 222 shows the torn out part of a blue print drawing of an Emery Wheel Arbor and gives the meaning of the Shaft marked with criss cross lines.

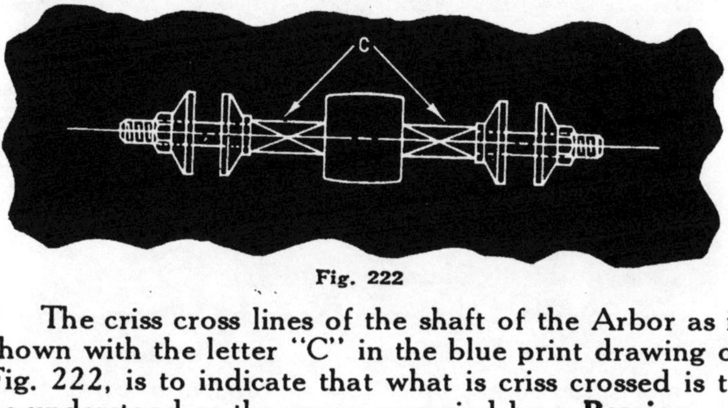

Fig. 222

The criss cross lines of the shaft of the Arbor as is shown with the letter "C" in the blue print drawing of Fig. 222, is to indicate that what is criss crossed is to be understood as the space occupied by a **Bearing.**

EFFECTS OF SHADING

The **effect of shading** is not really necessary in any blue print drawing, but its use, if properly drawn, helps to make the meaning clearer, besides placing an artistic touch to whatever is **shaded.** The theory of

Fig. 223

shading is the effect of lights and shadows cast upon an article to show its perspective.

Shading is most forcefully shown in round surfaces. Flat surfaces as shown in the end of Fig. 223 may be **shaded** to give the necessary contrast, showing that the surface is flat.

Figure 223 shows the round part of the Shaft spaced with broken lines, as the light **projects** from the equal imaginary spacings of its end, while the flat surface of the end is **shaded** to show the lights cast on its end.

The only real practical place to show **shading** is in an assembly blue print drawing, because of the artistic value it gives the drawing of the complete article, for **shading** always adds considerable weight for its acceptance by those concerned.

It is always a poor practice for **shading** to be shown in a detail blue print drawing, because **shading lines** confuse dimensions, while in assembly blue print drawings, dimensions are very rarely ever shown, and whatever dimensions are shown, would be only for the principal dimensions, such as, over all, center to center dimensions, etc.

HOW A BLUE PRINT DRAWING IS MADE

A blue print drawing is made on the same principle as a photograph is printed.

A photograph is printed from a plate or film taken by a camera, while a blue print drawing is printed from a **tracing cloth drawing** made by a draftsman, and such a **tracing cloth drawing** is much like a **film** for results obtained.

A **tracing** cloth drawing is cloth prepared with a glazed surface to hold ink, and its transparency enables the draftsman to draw his design, dimensions, remarks, etc., with ink from over the original rough pencil drawing on paper. Such a **tracing** cloth **drawing** is usually referred to as a **tracing.**

A **tracing** is as flexible and transparent as a **film,** and a blue print drawing is printed in the same manner as a photograph, that is why a blue print drawing is called a **"blue print."**

Blue print drawings are almost always printed on paper because of its cheapness, although in cases where there is hard usage of said blue print drawings, they may be printed on blue print cloth, which is more expensive.

Before a blue print drawing can be made, a sheet of dry paper or cloth that has been previously prepared with chemicals, white to all appearances when no strong light is about, is placed under the **tracing** and firmly held together under a plate of glass which acts as a clamp when secured in a frame. It is then exposed to a strong light for a certain length of time.

As the light penetrates thru the transparency of the **tracing,** it turns the chemically prepared paper underneath to be blue, except the places where there are ink lines or letters on the **tracing,** which will remain white on account of the light not being able to penetrate thru such black ink lines.

When such chemically prepared paper has been exposed long enough to a strong light to turn the paper blue wherever there is no ink lines, then such paper is taken out from underneath the **tracing** and placed in water to wash away the chemicals, especially where the ink lines of the **tracing** did not let the light turn the paper blue.

After the paper has been thoroughly washed of all chemicals, it is then hung up to dry and is soon ready for use.

Thru repeated operations, as many blue print drawings copies may be obtained as are desired.

A **tracing** offers a permanent record, and with the considerable amount of labor placed on the same, if properly handled, should last a life time. The permanent **tracing** would be closely guarded from fire or theft.

Blue print drawings for tool and other designs that are practically used but once, are drawn with pencil on a **tracing paper** called VELLUM, and blue prints printed from same, show faint white lines on the blue print drawing, on account of the transparency that allowed the light to penetrate thru the pencil lines of the **vellum tracing.**

CONCLUSION

In conclusion, it is hoped that this work will fulfill its aim, that it will explain "HOW TO UNDERSTAND BLUE PRINT DRAWINGS."

It is urged upon you to re-read this book every few weeks, until the fundamental principles stand out in your mind as never to be forgotten. Master the principles so thoroughly that whenever any kind of a blue print drawing is at hand, no matter how complicated, a complete understanding can be had with but little study.

Suggestions and criticisms for the betterment of this book will be welcome, so that the next edition may be strengthened to meet every demand of those who are working for skill in the reading of blue print drawings.

Spotted something that isn't quite right?

Let us know and we'll fix it for future reprints. And as a thank you for your diligence, we'll send you a free Sothis book of your choice.

publisher@sothis.press

This page intentionally left blank

Except of course for that line above. And this one.
And whilst we have your attention, perhaps you'd
like to follow us on Instagram @sothis.press. Go on.

Made in the USA
Monee, IL
05 November 2026

67807965R00102